DECLINE OF THE SEA TURTLES

DECLINE OF THE SEA TURTLES

CAUSES AND PREVENTION

Committee on Sea Turtle Conservation

Board on Environmental Studies and Toxicology

Board on Biology

Commission on Life Sciences

National Research Council

NATIONAL ACADEMY PRESS

Washington, D.C. 1990

NATIONAL ACADEMY PRESS 2101 Constitution Avenue, N.W. Washington, D.C.

NOTICE: The project that is the subject of this report was approved by the Governing Board of the National Research Council, whose members are drawn from the councils of the National Academy of Sciences, the National Academy of Engineering, and the Institute of Medicine. The members of the committee responsible for the report were chosen for their special competences and with regard for appropriate balance.

This report has been reviewed by a group other than the authors according to procedures approved by a Report Review Committee consisting of members of the National Academy of Sciences, the National Academy of Engineering, and the Institute of Medicine.

The project was supported by the Department of Commerce under contract number 50 DGNC 9 00080.

Cover: Loggerhead turtle. Photograph courtesy of the Florida Audubon Society.

Frontispiece: Thousands of fish and a loggerhead turtle caught in a shrimp trawl. The turtle is alive and apparently uninjured. The fish are dead. Photograph: Michael Weber, Center for Marine Conservation.

Photographs on chapter-opening pages are courtesy of Peter Pritchard and the Florida Audubon Society, and Michael Weber, Center for Marine Conservation.

Library of Congress Cataloging-in-Publication Data

Decline of the sea turtles: causes and prevention / Committee on Sea Turtle Conservation, Board on Environmental Studies and Toxicology, Board on Biology, Commission on Life Sciences, National Research Council.
p. cm.
Includes bibliographical references and index.
ISBN 0-309-04247-X: $14.95
1. Sea turtles. 2. Endangered species. 3. Wildlife conservation
I. National Research Council (U.S.). Committee on Sea Turtle Conservation.
QL666.C536D43 1990
639.9'7792--dc20

90-38318
CIP

Printed in the United States of America

Committee on Sea Turtle Conservation

John J. Magnuson, *Chair,* University of Wisconsin, Madison
Karen A. Bjorndal, University of Florida, Gainesville
William D. DuPaul, Virginia Institute of Marine Science, College of William and Mary, Gloucester Point
Gary L. Graham, Texas A&M University, College Station
David W. Owens, Texas A&M University, College Station
Charles H. Peterson, University of North Carolina, Morehead City
Peter C. H. Pritchard, Florida Audubon Society, Maitland
James I. Richardson, University of Georgia, Athens
Gary E. Saul, FTN Associates, Austin, Texas
Charles W. West, Nor'Eastern Trawl Systems, Bainbridge Island, Washington

Project Staff

David J. Policansky, Program Director
Dave Johnston, Project Director
Norman Grossblatt, Editor
Bernidean Williams, Information Specialist
Linda B. Kegley, Project Assistant

Board on Environmental Studies and Toxicology

Board on Biology

Commission on Life Sciences

Preface

The Committee on Sea Turtle Conservation was formed on March 1, 1989, under the auspices of the Board on Environmental Studies and Toxicology (BEST) and the Board on Biology of the National Research Council's Commission on Life Sciences. The committee was formally charged as follows:

> The task of this committee is to perform a study mandated by the Endangered Species Act Amendments of 1988, reviewing scientific and technical information pertaining to the conservation of sea turtles and the causes and significance of turtle mortality, including that caused by commercial trawling. The committee's report will provide information on the population biology, ecology, and behavior of five endangered or threatened species: the Kemp's ridley, loggerhead, leatherback, hawksbill, and green sea turtles. The committee will also review information on the effectiveness of current and needed programs to increase turtle populations. The resulting report will be used by the Secretary of Commerce to assess the effectiveness of and need for regulations requiring the use of turtle-excluder devices (TEDs) by commercial shrimp-trawlers.

In addition to this final report, the committee was required to prepare an interim report on the status of the Kemp's ridley (see Section 1008 of the Endangered Species Act Amendments of 1988 in Appendix A). The interim report on Kemp's ridley is enclosed as Appendix B.

The committee met on May 4-5, June 26-29, September 28-30, and November 16-18, 1989. In addition, a writing group met in Washington, D.C., on December 4-5, 1989.

The committee was fortunate to have as members experts on turtle biology, physiology, and conservation, shrimp fishing and gear technology (in particular, trawl and TED technology), fishery biology and management, and technology transfer and fishery extension programs. All these kinds of expertise were required to address the complex issues of sea turtle conservation. I was especially pleased by the new analyses and syntheses that the committee was able to make that allowed us to reach a number of important conclusions with more certainty than I originally had expected.

To gain further knowledge of TEDs and other gear operations, the committee sailed for a day off the coast of Jekyll Island, Georgia, on the *RV/Georgia Bulldog*. The trip was made possible through the courtesy of David Harrington.

We were assisted greatly by the National Research Council staff, in particular, Dave Johnston, our project director, and Linda Kegley, our project assistant. We also acknowledge the guidance of David Policansky, program director; James J. Reisa, director of BEST; and Joanna Burger, our BEST liaison. In addition to those persons, other BEST staff jumped in to help when needed. Bernidean Williams, information specialist, helped to check references. Norman Grossblatt of the Commission on Life Sciences and Lee Paulson of BEST edited the report. My personal thanks to them all.

We also thank a number of persons who either prepared presentations for the committee or made original data available to the committee for analysis and interpretation. Those who provided testimony were Michael Weber (Center for Marine Conservation), Larry Ogren (National Marine Fisheries Service), Earl Possardt (U.S. Fish and Wildlife Service), Charles Oravetz (National Marine Fisheries Service), Tee John Mialjevich (Concerned Shrimpers of America), Carol Ruckdeschel (Cumberland Museum), David Harrington (University of Georgia Marine Extension Service), Joe Webster (Commercial Shrimp Fisherman), David Blouin (Department of Experimental Statistics, Louisiana State University), Elizabeth Gardner (legislative assistant to Senator Howell Heflin), and David Cottingham (National Marine Fisheries Service). Those who provided data include Wendy Teas, Nancy Thompson, Terry Henwood, Warren Stuntz, Edward F. Klima, Ren Lohoefener, Ernie Snell, James Nance, and Guy Davenport, all with the National Marine Fisheries Service; Alan Bolten (University of Florida); Michael Harris (Georgia Department of Natural Resources); Tom Henson (North Carolina Nongame Endangered Wildlife Program); and Sally Murphy (South Carolina Wildlife and Marine Resources). Without their assistance we would not have been able to make our report current.

I wish to thank my own colleagues in the Center for Limnology at the University of Wisconsin-Madison, who assisted with analyses and data transfer: Chris Carr, Barbara J. Benson, Inga Larson, Norma D. Magnuson, Mark D. McKenzie, Mary Rose Lawecki Smith, and Joyce M. Tynan. Assistance to other committee members was provided by Nancy Balcom (Virginia Institute of Marine Science), David Rostal (Texas A&M University), and Hal Summerson and Alberto Reyes-Campo (University of North Carolina, Institute of Marine Sciences).

We dedicate this volume to the peaceful coexistence of sea turtles and shrimp fisheries.

John J. Magnuson
Chairman
April 23, 1990

Contents

Figures and Tables

Executive Summary

Five species of sea turtles regularly spend part of their lives in U.S. coastal waters of the Atlantic Ocean and the Gulf of Mexico: Kemp's ridley, loggerhead, green turtle, hawksbill, and leatherback. They are ancient reptiles, having appeared on earth millions of years before humans. Sea turtles were widely used by humans in earlier times for food, ornaments, and leather, and they still are used in these ways by many societies. They are now endangered or threatened and are protected under the Endangered Species Act. Kemp's ridleys, leatherbacks, and hawksbills are listed as endangered throughout their ranges; green turtles are endangered in Florida, and threatened in all other locations; loggerheads are listed as threatened throughout their range. For some major populations and species of sea turtles to persist, substantial progress in conservation will have to be made.

Concerns about the continuing declines of sea turtle populations and the potential impact of new gear regulations on commercial shrimp trawlers prompted the Congress to add a provision to the Endangered Species Act Amendments of 1988 mandating an independent review by the National Academy of Sciences of scientific and technical information pertaining to the conservation of sea turtles. The Congress further mandated review of the causes and significance of turtle mortality, including that caused by commercial trawling. Accordingly, a study committee was convened by the National Research Council's Board on Environmental Studies and Toxicology in collaboration with its Board on Biology. The committee included experts in international and domestic sea turtle biology and ecology, coastal zone development and management, commercial fisheries and gear technology, marine resources, and conservation biology. During the course of the committee's 1-year study, it heard from representatives of the shrimping industry, conservation organizations, the U.S. Fish and Wildlife Service, the National Marine Fisheries Service, and Sea Grant programs. The committee observed shrimp trawling exercises with and without turtle excluder devices on a converted shrimp trawler in Georgia coastal waters. It reviewed pertinent published literature and analyzed original data sets on aerial and beach turtle surveys, shrimp trawling efforts, other commercial fisheries, turtle strandings, and other materials from a variety of organizations and knowledgeable individuals.

This report presents scientific and technical information on the population biology, ecology, and reproductive behavior of five endangered or threatened species of sea turtles. It evaluates population declines, causes of turtle mortality, and the effectiveness of past and current mitigation efforts, and recommends conservation measures to protect or increase turtle populations. The committee was not charged or constituted to address and did not analyze social and economic issues related to sea turtle conservation.

LIFE HISTORIES OF SEA TURTLES

The five species of sea turtles considered in this report have similar life histories. Females of all five species lay clutches of about 100 eggs and bury them in nests on coastal beaches. Mature male and female sea turtles aggregate off the nesting beaches during the spring to mate, and females might return to the beach to deposit 1 to 10 clutches in a season. Individual Kemp's ridleys probably nest each year after reaching maturity; females of the other species routinely nest every 2-4 years.

After an incubation period of about 2 months, hatchlings of all the species dig their way to the surface of the sand and scramble over the beach in their short trip to the ocean. Once in the water, they swim offshore and spend their early life near the surface in the offshore waters of the Atlantic or Gulf of Mexico. After a few years, most species enter the coastal zone or move into the bays, river mouths, and estuaries, where they spend their juvenile life, eating and growing until they reach maturity some 10-50 years later. Mature sea turtles usually weigh 35-500 kg.

Food habits differ among species. Kemp's ridleys prefer crabs, loggerheads eat a wide range of bottom-dwelling invertebrates, green turtles eat bottom-dwelling plants, leatherbacks prey on jellyfish in mid-water, and hawksbills specialize on bottom-dwelling sponges.

SEA TURTLE DISTRIBUTION AND ABUNDANCE

Judged from strandings of carcasses on beaches from the Mexican border to Maine, the most abundant sea turtles in U.S. coastal waters are loggerheads, followed by Kemp's ridleys, green turtles, leatherbacks, and hawksbills. According to aerial surveys, large loggerheads are most abundant off the coasts, and leatherbacks are about one-hundredth as abundant as loggerheads in the Atlantic. In general, other adult turtles and smaller juveniles are difficult to see and identify from the air.

One of the two largest loggerhead rookeries in the world is concentrated along the Atlantic beaches of central and southern Florida, but loggerheads nest from southern Virginia to eastern Louisiana. Aerial surveys have identified large concentrations of loggerheads off their primary nesting beaches in Florida during the spring and summer; sightings off the nesting beaches are much less frequent during the autumn and winter.

Regular nesting of green turtles and leatherbacks also occurs on the Atlantic beaches of central and southern Florida. Kemp's ridleys and hawksbills do not make important use of U.S. coastal beaches, except for hawksbills in the U.S. Caribbean islands.

Based on limited trawling data in the gulf, juvenile and adult sea turtles off the South Atlantic and gulf coasts are more abundant in waters less than 27 m deep than in deeper waters. Limited aerial surveys in the gulf reveal they are more abundant in waters less than 50 m. Data on depth distribution are scarce, but turtle density during shrimping seasons is apparently about 10 times greater in shallow than in deeper waters.

SEA TURTLE POPULATION TRENDS

Changes in sea turtle populations are most reliably indicated by changes in the numbers of nests and nesting females on the nesting beaches. Females return to the same beaches repeatedly and are relatively easily counted there. For trend analysis, the incidence of carcass strandings on the beaches and the number of adults sighted at sea from airplanes are much less satisfactory, because of uncontrolled variables and uncertainties.

The results of population-trend studies are clear in several important cases. Kemp's ridley nesting populations have declined to about 1% of their abundance in 1947 at their only important nesting beach, Rancho

Nuevo, on the Mexican coast of the Gulf of Mexico. Since 1978, the number of Kemp's ridley nests has been declining at about 14 per year; the total number of nesting females currently might be as low as 350 (although clearly there are additional turtles in the population: juveniles and males). Loggerhead populations nesting in South Carolina and Georgia are declining, but populations on parts of Florida's Melbourne Beach and Hutchinson Island apparently are not declining, and the Hutchinson Island population might even be increasing. Green turtles nesting on Hutchinson Island are increasing. Data are insufficient to determine whether other populations in U.S. waters are increasing or decreasing. Data available on hawksbills or leatherbacks do not show clear-cut trends in U.S. waters.

NATURAL MORTALITY OF SEA TURTLES AND REPRODUCTIVE VALUE OF LIFE STAGES

Mature female sea turtles lay many clutches of eggs during their lifetimes with about 100 eggs per clutch, but only about 85% of the undisturbed eggs produce hatchlings, and most of the hatchlings probably die in their first year. The greatest source of natural mortality of these eggs and hatchlings is predation, primarily by carnivorous mammals, birds, and crabs in and on the beaches and by birds and predatory fishes in the ocean. Shoreline erosion of dunes and inundation (drowning) of nests are other important sources of natural mortality. Various causes of sea turtle mortality associated with human activities (artificial lighting, coastal development, etc.) are usually an important component of total mortality. As juvenile turtles in the shallow coastal zone reach a larger size (58-79 cm long), natural mortality rates are expected to decline. A female loggerhead probably reaches maturity at about 20-25 years, remains reproductively active for another 30 years or so, and produces a very large number of eggs during her lifetime.

The consideration of age-specific natural mortality and reproduction leads to the important concept of reproductive value for each of a turtle's life stages. Reproductive value is a measure of how much an individual at a particular stage of life contributes to the future growth or maintenance of the population. An analysis of reproductive value provides valuable insight for decision makers responsible for the conservation of sea turtles, because it indicates which individuals contribute most to future populations and also where protection is likely to be the most effective. One life-stage analysis of reproductive value for eggs and hatchlings, small juveniles, large juveniles, subadults, and nesting adults used loggerheads at Little Cumberland Island, Georgia, as the example. It was concluded that the key to improving the outlook for Georgia and Carolina nesting loggerhead populations lies in reducing the mortality in the older stages, particularly the large juveniles 58-79 cm long. Because the reproductive value of the earliest stage was so very low compared with the older stages, protecting 100% of the eggs and hatchlings was not sufficient to reverse the decline in the numbers of nesting females of this model population. It was also noted that the 58-79 cm group of large juveniles is the size class that dominates in the distribution of stranded carcasses on beaches from northern Florida to North Carolina.

The committee concluded that conservation measures directed at large juveniles and adults are especially critical to the success of sea turtle conservation.

SEA TURTLE MORTALITY ASSOCIATED WITH HUMAN ACTIVITIES

All life stages of sea turtles are susceptible to human-induced mortality. Direct human manipulations—such as beach armoring, beach nourishment, beach lighting, and beach cleaning—can reduce the survival of eggs and hatchlings in and on the beaches. The presence of humans on the beach, on foot or in vehicles, can adversely affect nesting, buried eggs, and emerging hatchlings. Other factors, such as beach erosion and accretion, or the introduction of exotic plants and predators, are indirect effects of humans that can be responsible for many turtle deaths.

However, the committee's analyses led it to conclude that for juveniles, subadults, and breeders in the coastal waters, the most important human-associated source of mortality is incidental capture in shrimp trawls, which accounts for more deaths than all other human activities combined. The committee estimated that mortality from shrimping lies between 5,000-50,000 loggerheads and 500-5,000 Kemp's ridleys each year. Collectively, other trawl fisheries; fisheries that use passive gear, such as traps, gill nets, and long lines; and entanglement in lost or discarded fishing gear and debris are responsible for an additional 500-5,000 loggerhead deaths and 50-500 Kemp's ridley deaths a year. Although those numbers are an order of magnitude lower than the losses due to the shrimp fisheries, they are important. Next in importance are the deaths due to dredging, and collisions with boats: an estimated 50-500 loggerheads each and 5-50 Kemp's ridleys each. Oil-rig removal could account for 10-100 turtle deaths per year, and deaths from intentional harvest of turtles in U.S. coastal waters and entrainment by electric power plants are judged each to be fewer than 50 per year. Deaths resulting from ingestion of plastics and debris and from accumulation of toxic substances, especially from ingested petroleum residues, could be important, but the committee was unable to quantify them.

The estimates of human-associated sea turtle deaths are most certain for shrimp fishing and power-plant entrainment; they are less certain for dredging, and least certain for other fisheries, collisions, oil-rig removal, intentional harvest, and ingestion of plastics or debris. In some cases, although direct estimation is impossible, worst-case estimates provide an upper limit on the potential mortality associated with oil-rig removal and collisions with boats. In some cases, conservation measures are in place or are being implemented, and these will lower the above estimates.

The Shrimp Fishery

The U.S. shrimp fishery is a complex of fisheries from Cape Hatteras, North Carolina, to the Mexican border in the gulf. Those fisheries harvest various species of shrimp at various stages in their life cycles, using a variety of vessels that range from ocean-going trawlers to small vessels operating in nearshore or inside waters. About one-third of the shrimping effort occurs in bays, rivers, and estuaries; two-thirds occurs outside the coastline. Ninety-two percent of the total effort is in the gulf; most of that is in waters shallower than 27 m. The fishing areas off the coastal beaches of Texas and Louisiana account for 55% of the total U.S. effort and 83% of the effort off the coastal beaches. In the Atlantic, 92% is within 5 km of shore. One important nesting area for turtles, where almost no shrimping effort occurs, is the central to southern portion of the Atlantic coast of Florida. Atlantic shrimping effort is concentrated off South Carolina, Georgia, and northern Florida.

Several lines of strong evidence make it clear that sea turtle mortality due to incidental capture in shrimp trawls is large:

- The proportion of dead and comatose turtles in shrimp trawls increases with tow time of the trawl—from very few at 40 minutes to about 70% after 90 minutes.
- The number of stranded carcasses on the beaches increases stepwise by factors of 3.9 to 5 when shrimp fisheries open in South Carolina and Texas, and decreases stepwise when a shrimp fishery closes in Texas. The data suggest that 70-80% of the turtles stranded at those times and places were caught and killed in shrimp trawls.

- Loggerhead nesting populations are declining in Georgia and South Carolina, where shrimp fishing is intense, but are not declining and might even be increasing farther south in central and southern Florida, where shrimp fishing is rare or absent. The committee is aware that these interactions are complex.
- A much-cited estimate of shrimping-related mortality, 11,000 loggerheads and Kemp's ridleys per year in U.S. coastal waters of the Atlantic and the gulf, was judged by this committee to be an underestimate, possibly by as much as a factor of 4. This maximal value of 44,000 falls within the order of magnitude estimates by the committee that the number of loggerheads and Kemp's ridleys killed annually lies between 5,500 and 55,000. The estimate of 11,000 turtles killed annually was based on analysis that did not account for mortality in bays, rivers, and estuaries, even though many turtles and one-third of the shrimping effort occurs there. The estimate was also based on the assumption that all comatose turtles brought up in shrimp nets would survive. Recent observations have suggested that many (perhaps most) comatose turtles will die and should be included in the mortality estimates until effective rehabilitation methods are available and used.
- In North Carolina, turtle stranding rates increase in the summer south of Cape Hatteras while the shrimp fishery is active there, and in the fall and winter north of Cape Hatteras while the flounder trawl fishery is active there. That observation suggests that the flounder fishery might be another source of mortality north of the cape in the fall and winter.

Other Fisheries

Mortality associated with other fisheries and with lost or discarded fishing gear is much more difficult to estimate than that associated with shrimp trawling, and there is a need to improve the estimates. A few cases stand out, such as the possible turtle losses from the winter flounder trawl fishery north of Cape Hatteras (about 50-200 turtles per year); the historical Atlantic sturgeon fishery, now closed, off the Carolinas (about 200 to 800 turtles per year); and the Chesapeake Bay passive-gear fisheries (about 25 turtles per year). Considering the large numbers of fisheries from Maine to Texas that have not been evaluated and the problems of estimating the numbers of turtles entangled in the 135,000 metric tons of plastic nets, lines, and buoys lost or discarded annually, it seems likely that more than 500 loggerheads and 50 Kemp's ridleys are killed annually by nonshrimp fisheries.

Dredging

Estimates of the mortality of sea turtles taken in dredging operations range from 0.001 to 0.1 per hour. If it takes 1,000 hours of dredging to maintain each navigation channel each year, one to 100 turtles could be killed per active channel in areas frequented by turtles. The 0.1 per hour might be an unrealistically high estimate, and some conservation measures are in place, so the number of turtles killed per channel is probably much less than 100 per year.

Boat Collisions

Boat collisions with turtles are evident from damage to turtles that strand on coastal beaches. Many of them could have been dead before they were hit, but not all turtles hit and killed by boats drift ashore. The committee estimates that a maximum of 400 turtles per year are killed by collisions off the coasts, but the estimate is very uncertain and unknown for inside waters.

Oil Platforms

About 100 oil platforms in the western gulf are scheduled for removal each year for the next 10 years. The probability of there being at least one turtle within the damage zone (i.e., within 1,000 m of an explosion to remove a rig) is estimated to be between 0.08 and 0.50. That yields a minimal estimate of 8-50 turtle deaths per year. This estimate might be low, because it is based only on aerial sightings of turtles, or high, because rigs will be surveyed and attempts made to move turtles out of the region before rig removal.

Plastics and Debris

About 24,000 metric tons of plastic packaging is dumped into the ocean each year. The occurrence of plastic debris in the digestive tracts of sea turtles is common; for example, half the turtles that stranded on Texas beaches in 1986-1988 and one-third of the leatherbacks and one-fourth of the green turtles from the New York Bight area necropsied in 1979-1988 had plastic debris in their digestive tracts. The food preferences of the leatherback (jellyfish) and green turtle (bottom plants), in particular, could make them especially susceptible to ingestion of plastic bags. Ingestion of plastics could interfere with food passage, respiration, and buoyancy and could reduce the fitness of a turtle or kill it. Floating plastics and other debris, such as petroleum residues drifting on the sea surface, accumulate in sargassum drift lines commonly inhabited by hatchling sea turtles during their pelagic stage; these materials could be toxic.

The committee was unable to make quantitative estimates of mortality from these sources, but the impact of ingesting plastics or debris could be severe.

SEA TURTLE CONSERVATION

The committee considered conservation measures applicable to the two habitats of sea turtles most vulnerable to human-associated mortality: the beaches (eggs, hatchlings, and nesting females) and the coastal zone (juveniles, subadults, and breeders). The first set of conservation measures pertains to activities on the nesting beaches and to supplementing reproduction; the second, to activities in the coastal zone off the coastal beaches and in the bays, rivers, and estuaries.

Eggs, Hatchlings, and Nesting Females

Nesting Habitat

Critical nesting habitat can be protected through various types of public and private ownership and regulation of beach activities. Increased protection can prevent damage from beach armoring, beach nourishment, and human use, including vehicular traffic. Relocation of nests can also help, but must be done by qualified and approved groups. The disorientation caused by artificial lighting might be reduced with the use of low-pressure sodium lights. Some municipalities in Florida have passed lighting ordinances. Protection of eggs from predators and predator control on some beaches are important conservation measures. Kemp's ridley eggs at Rancho Nuevo still must be removed from the nests and protected from human and coyote predation to ensure their survival; almost all eggs are transferred to an enclosed beach hatchery and thus protected from predation.

Headstarting

Headstarting is an attempt to reduce the mortality of hatchlings by rearing them in captivity to a size at which their mortality rate in the wild should be lower. It is an active experiment with the Kemp's ridley, but headstarting has not yet proved to be effective. Benefits are uncertain, because some headstarted turtles appear to behave abnormally in the wild, many are soon caught in various fisheries, and none has yet been recorded as reaching maturity or nesting. Headstarting methods have improved greatly, and proponents argue that the experiment has not yet received a fair test. The program has research and public-awareness benefits. Regardless, headstarting cannot be effective without concurrent reduction in the mortality of juveniles in the coastal zone.

Captive Breeding

Loggerheads, green turtles, and Kemp's ridleys have been raised in captivity from eggs to adults. The same species lay fertile eggs in captivity. However, despite successes in captive breeding programs, the committee does not consider captive breeding to be a preferred management tool. If a species became extinct except for captive animals, it would probably not be feasible to re-establish the wild population from captive animals, because captive animals in an aquarium or zoo would retain only a portion of the genetic material of their species.

Artificial Imprinting

Some limited evidence suggests that hatchlings might imprint on their natal beaches. The extent to which artificial imprinting might promote new nesting sites or restore old ones remains uncertain.

Juveniles, Subadults, and Breeders

Conservation measures applicable to juveniles, subadults, and breeders involve the reduction of intentional harvest, reduction of unintentional capture and deaths in fishing gear, and modification of dredging operations, oil-rig removal, and various other sources of human-associated mortality.

Prohibition of Intentional Harvest

Intentional harvest of sea turtles in U.S. waters is prohibited by the Endangered Species Act. The increase in numbers of green turtles nesting at one site in southern Florida might be early evidence that prohibition has been effective. Similar protection has been implemented in Mexico, but enforcement is imperfect. Intentional harvest of sea turtles and their eggs continues to occur throughout the Caribbean region, including Puerto Rico.

Reduction of Unintentional Bycatch

Sea turtle deaths caused by unintentional capture in shellfish and finfish fisheries can be reduced by limiting fishing effort at some times and places, closing a fishery, modifying fishing gear to exclude turtles or, for trawl fisheries, reducing the tow times. New technology, such as the use of turtle excluder devices (TEDs) in bottom trawls and smaller mesh size in pound-net leaders, can reduce turtle deaths.

Fishery closures can be effective, as demonstrated in the case of the sturgeon fishery off the Carolinas and as evidenced by the maintenance of sea turtle nesting rookeries in the south Atlantic coast of Florida, where there is very little shrimp fishing. There might be some areas and seasons in which turtles are so common that a fishery should be closed and other areas and seasons in which turtles are so uncommon that fishing could occur without the need for devices or procedures to reduce turtle mortality. One area to consider for less stringent measures to prevent turtle deaths is the deeper waters of the Gulf of Mexico. Distribution data should be examined in detail to locate possible sites on fine spatial and temporal scales, for example by month, fishing zone, and depth.

Turtle excluder devices are designed for installation in shrimp-trawling gear to release turtles from the net without releasing shrimp. By November 1989, six TED designs had been shown to exclude 97% of the sea turtles that would have been caught in nets without TEDs. They have been certified by the National Marine Fisheries Service to exclude turtles. Some, such as the Georgia jumper, have stiff frames; others, such as the Morrison soft TED, are made only of soft webbing. The various designs differ in their ability to retain shrimp. Under good conditions, some designs have not been shown to reduce shrimp catch, whereas others have. A TED's performance also is affected by the roughness of the bottom and the amount of debris or vegetation on the bottom. Debris can collect on a TED and degrade the efficiency of the TED in excluding turtles and the efficiency of the net in capturing shrimp. Reduction of tow time might be a preferable alternative to the use of TEDs in some locations if there is too much debris. In some situations, a TED can improve the efficiency of trawling by excluding cannonball jellyfish, which otherwise would clog the net.

Fishing effectively with TEDs requires some skill in adapting to local situations, but overall it is an effective way to protect the juveniles and adults that are important to the maintenance and recovery of sea turtle populations. TED technology transfer is crucial, because TEDs are effective in excluding turtles from shrimp trawls. The National Marine Fisheries Service has relied heavily on the Sea Grant program to help in the transfer of TED technology to shrimp fleets. Many activities have been undertaken, such as workshops, hearings, dockside and on-board demonstrations, presentations at industry meetings, and distribution of a large variety of written information. But the responses of commercial shrimpers to these initiatives have been poor in many areas.

Making tow times shorter than those which kill turtles might work in some situations in which short tow times are feasible. If tow times are limited to 40 minutes in the summer and 60 minutes in the winter, few, if any, captured turtles die or become comatose. Comatose turtles should be counted as dead, until effective rehabilitation techniques for comatose turtles can be developed and demonstrated. Limiting tow time is probably more feasible with small boats in shallow waters. Even so, the problem of multiple successive recaptures must be solved.

Dredging

With respect to dredging, conservation measures might have included relocation, but in trials, some turtles have returned to the dredging area after an unacceptably short time. Several actions have been initiated: putting observers on dredges, comparing different dredge designs, redesigning deflectors, and studying the behavior and distribution of sea turtles in key navigation channels. Studies of the latter type in the Port Canaveral Entrance Channel have led to restricting dredging to the fall, when turtles are least abundant there.

Collisions with Boats

Collisions of boats with turtles are difficult to count, and conservation measures are inherently difficult to implement. Better evaluation of the extent of the problem could lead to production and distribution of educational material and some boating rules in inside waters with high concentrations of turtles.

Oil-Rig Removal

The impact of oil-rig removal on sea turtles is poorly documented. Conservation measures should include surveys and removal of sea turtles before oil-rig demolition and further evaluation of the extent of the problem.

Power Plants

A few sea turtles are still being entrained at the intake pipes of some power plants. Use of tended barrier nets to remove sea turtles could reduce this small source of mortality.

Plastics and Debris

The best conservation measures to reduce ingestion of plastics and debris are measures that reduce ocean dumping of such materials from ships and land sources. The International Convention for the Prevention of Pollution from Ships (known as MARPOL) makes it illegal to dispose of any plastics at sea. It also sets down guidelines to prohibit dumping of garbage (of the galley type) in nearshore waters. The consequences for sea turtles of ingesting plastics and debris are poorly understood, and the subject needs further study.

Education

Public education is important for calling attention to sea turtle conservation and implementing the conservation measures. Good beach management stems from an informed and educated public. Many published materials are already available, and others will be needed, especially on the effects of fisheries on the sea turtle life stages with the highest reproductive value and on the effects of ingesting plastics and other debris.

Research

Research projects on sea turtles have been many and varied, and they span such broad categories as distribution, population trends, food habits, growth and physiology, and major threats to survival. The committee recognizes the need to improve the data bases for each of those categories, to establish long-term surveys of sea turtle populations at sea and on land, and to initiate experimental programs to increase population sizes.

CONCLUSIONS AND RECOMMENDATIONS

Conclusions

1. Combined annual counts of nests and nesting females indicate that nesting sea turtles continue to experience population declines in most of the United States. Declines of Kemp's ridleys on the nesting beach in Mexico and of loggerheads on South Carolina and Georgia nesting beaches are especially clear.
2. Natural mortality factors—such as predation, parasitism, diseases, and environmental changes—are largely unquantified, so their respective impacts on sea turtle populations remain unclear.
3. Sea turtles can be killed by several human activities, including the effects of beach manipulations on eggs and hatchlings and several phenomena that affect juveniles and adults at sea: collisions with boats, entrapment in fishing nets and other gear, dredging, oil-rig removal, power plant entrainment, ingestion of plastics and toxic substances, and incidental capture in shrimp trawls.
4. The incidental capture of sea turtles in shrimp trawls was identified by this committee as the major cause of mortality associated with human

activities; it kills more sea turtles than all other human activities combined.
5. Shrimping can be compatible with the conservation of sea turtles if adequate controls are placed on trawling activities, especially the mandatory use of turtle excluder devices (TEDs) at most places at most times of the year.
6. The increased use of conservation measures on a worldwide basis would help to conserve sea turtles.

Recommendations

1. Trawl-related mortality must be reduced to conserve sea turtle populations, especially loggerheads and Kemp's ridleys. The best method currently available (short of preventing trawling) is the use of TEDs. Therefore, although the waters off northern Florida, Georgia, South Carolina, Louisiana, Mississippi, Alabama, and Texas are most critical, the committee recommends the use of TEDs in bottom trawls at most places and most times of the year from Cape Hatteras to the Texas-Mexico border. At the few places and times where TEDs might be ineffective (e.g., where there is a great deal of debris), alternative conservation measures for shrimp trawling might include tow-time regulations under very specific controls, and area and time closures, as discussed in Chapter 7. Available data suggest that limiting tow times to 40 minutes in summer and 60 minutes in winter would yield sea turtle survival rates that approximate those required for approval of a new TED design. Restrictions could be relaxed where turtles are and historically have been rare.
2. Conservation and recovery measures for all sea turtle species that occur in U.S. territorial waters should include protection of nesting habitats, eggs, and animals of all sizes. Of special concern are the nesting beaches of Kemp's ridleys in Mexico and of loggerheads between Melbourne Beach and Hutchinson Island in Florida. Undeveloped beach property between Melbourne Beach and Wabasso Beach, Florida, in the Archie Carr National Wildlife Refuge proposed by the U.S. Fish and Wildlife Service, should be protected. Lands are available for purchase, and action should be taken now.
3. Incidental deaths associated with other human activities—such as other fisheries and abandoned fishing gear, dredging, and oil-rig removal—should also be addressed and reduced.
4. Headstarting should be maintained as a research tool, but it cannot substitute for other essential conservation measures.
5. Research on sea turtles should include improvement of the data base

on survivorship, fecundity, mortality at all life stages; distribution and movements; effects of ingesting plastics and petroleum particles; parasitism and disease, and other pathological conditions; and physiology of sea turtles, especially their resistance to prolonged submergence and their recovery from a comatose condition. Carefully designed and implemented long-term surveys of sea turtle populations both on land and in the sea will be crucial to their survival. The cumulative effects of human activities on nesting beaches should be quantified relative to the total available nesting areas, because the loss of nesting beaches through development or alteration could extirpate local populations.

6. Efforts to improve TED technology and explore other methods to conserve sea turtles should be continued, including research on the effectiveness of regulations.

1

Introduction

All species of sea turtles that live in U.S. waters are listed as endangered or threatened under the Endangered Species Act of 1973 (ESA). An endangered species is one that is in danger of extinction throughout all or a significant portion of its range; a threatened species is one that is likely to become endangered throughout all or a significant portion of its range within the foreseeable future. The ESA requires protection of both categories. A principal goal of this report, which was mandated by the 1988 amendments to the ESA, is to provide a sound scientific basis for protecting these endangered and threatened species of sea turtles.

The leatherback (*Dermochelys coriacea*) and hawksbill (*Eretmochelys imbricata*) were listed as endangered throughout their ranges on June 2, 1970. The Kemp's ridley (*Lepidochelys kempi*) was listed as endangered on December 2, 1970. The green turtle (*Chelonia mydas*) was listed on July 28, 1978, as threatened, except for the breeding populations of Florida and the Pacific coast of Mexico, which were listed as endangered. On July 28, 1978, the loggerhead (*Caretta caretta*) was listed as threatened throughout its range.

Those sea turtles were listed because, to different degrees, their populations had declined largely as a result of human activities. They have been prized worldwide as meat for human consumption, their eggs consumed or used as aphrodisiacs, their oil used for lubricants and ingredients in cosmetics, and their shells used for jewelry and eyeglass frames. Mass slaughter of turtles and plunder of their nests have been and remain a prime cause of population declines. Many nesting beaches were severely degraded by encroachment of human populations into coastal habitats. Sea turtle populations have been reduced by uncontrolled harvesting for personal or commercial purposes and by mortality incidental to such activities as commercial fishing.

For at least 2 decades, however, several factors appear to have contributed unevenly but increasingly to the decline of sea turtle populations along the Atlantic coast and in the Gulf of Mexico: physical and ecological degradation of turtle nesting habitats; plastics and persistent debris in marine ecosystems; continued turtle harvesting in international waters; activities associated with oil and gas development; collision with power boats; explosive devices; and shrimp trawling. In fact, several reports in the 1980s argued that the inadvertent capture and mortality (presumably through drowning) of sea turtles in shrimp trawls were major factors hindering the recovery of the species.

The ESA prohibits capture of endangered sea turtles within the United States and its territorial waters and on the high seas, except as authorized by the Secretary of Commerce or the Secretary of the Interior. The Secretary of Commerce has authority over sea turtles in marine waters and the Secretary of the Interior has authority over sea turtles on land. ESA authorizes the secretaries to extend to threatened species the same protections provided to endangered species. Under the ESA, it is unlawful to import, export, take, possess, sell, or transport endangered species without a permit, unless these activities are specifically allowed by regulation.

Early observations of sea turtle populations strongly indicated that inadvertent capture and death of sea turtles in shrimp trawls was a major mortality factor of the species. To prevent further declines in the populations of the five species of sea turtles, the National Marine Fisheries Service (NMFS) in about 1978 began to develop research and public-education programs aimed at decreasing sea turtle mortality in the Gulf of Mexico and southern Atlantic states. Guidelines for resuscitating and releasing turtles incidentally caught in their trawling operations were developed by NMFS and the active participation of shrimpers. Gear-research programs under the auspices of NMFS, Sea Grant, and the shrimping industry itself led to the development of several types of net installation devices that came to be called turtle excluder devices (TEDs) or, later, trawler efficiency devices.

The only NMFS-approved TED in 1983 was an NMFS TED, and by early 1986, only certain versions of this device were approved. Because many fishermen were apprehensive about using TEDs in mid-1986, the

University of Georgia and NMFS tested industry designs at Cape Canaveral. This resulted in NMFS certification of the Georgia, Cameron, and Matagorda TEDs. In the summer of 1987, the Morrison TED was certified, and an early version of the Parrish TED was tested. In the fall of 1987, a modified Parrish TED was certified.

Each type of TED was intended to divert swimming turtles out of shrimp nets, thus excluding the turtles from the nets while not reducing the shrimp catch. Over about a decade of development, TEDs were lightened and modified from the prototype. Today, six kinds have been approved by NMFS for use on shrimp-trawling vessels (Appendix C). Each TED has been repeatedly tested for effectiveness by NMFS, state agencies, and private shrimpers.

By 1983, NMFS had tried a voluntary compliance program encouraging shrimpers to use TEDs, but few shrimpers responded. Instead, most shrimpers regarded TEDs as nuisances and remained unconvinced that the devices provided sufficient economic incentive in the form of catch purity. Some argued that TEDs would reduce their shrimp catches and that TEDs are expensive, dangerous, and time-consuming to install and clean—all this adding to the monetary costs of shrimping. Primarily because of results of testing different TEDs under different conditions, both NMFS and environmental groups became convinced that TEDs effectively exclude turtles from shrimp nets and that their use does not result in a significant reduction in the shrimp catch. In fact, field tests in different areas indicated that the best TEDs sometimes reduced the incidental catch of turtles by up to 97% with little or no loss in the shrimp catch.

A conflict arose almost immediately between proponents of TED regulations and the gulf shrimping industry. Shrimpers were not convinced that the turtles killed in shrimp trawls were responsible for the reported overall declines in sea turtle populations. They believed that something else was killing the turtles. Representatives of the industry in the gulf area categorically asserted that the imposition of TEDs on trawlers would reduce shrimp catch and devastate the industry. Several lawsuits were filed to delay the implementation of the NMFS regulations regarding TEDs.

By 1985, it was apparent that relatively few shrimpers were using TEDs voluntarily. Faced with the threat of lawsuits to close down the shrimping industry, NMFS sponsored a series of mediation meetings in 1986 that included members of environmental organizations and shrimpers (Conner, 1987). The group agreed (with one abstention) to a negotiated rule-making that would phase in the required use of certified TEDs in specific areas at specific times. By 1987, however, grassroots pressure led to state and federal legislative attempts to delay the implementation of TED regulations, and some of the industry parties to the negotiated rulemaking repudiated the agreement (Conner, 1987). After numerous debates, conferences, and public hearings over the years, NMFS developed by 1987 a final set of regulations on the use of TEDs by shrimp trawlers, to be implemented in 1989. The regulations—including trawler size, geographic zones, seasons, tow-time restrictions, exemptions, and starting dates—were published in the *Federal Register* [52 (124):24247-24262, June 28, 1987].

The controversy and concern over the deaths of turtles in trawl nets of shrimpers and the potential effects of the proposed regulations to protect turtles on the shrimping industry then motivated Congress to amend reauthorization of the Endangered Species Act in 1988. One of the amendments to the reauthorization stipulated that a committee of the National Academy of Sciences should review the biology and behavior of the five species of sea turtles.

Section 1008 of the Endangered Species Act Amendments of 1988 specified the following issues for study:

- Estimates of the status, size, age structure, and, where possible, sex structure of each of the relevant species of sea turtles.
- The distribution and concentration, in terms of United States geographic zones, of each of the relevant species of sea turtles.
- The distribution and concentration of each of the relevant species of sea turtles, in the waters of the United States, Mexico, and other nations during the developmental, migratory, and reproductive phases of their lives.
- Identification of all causes of mortality, in the waters and on the shores of the United States, Mexico, and other nations for each of the relevant species of sea turtles.
- Estimates of the magnitude and significance of each of the identified causes of turtle mortality.
- Estimates of the magnitude and significance of present and needed headstart or other programs designed to increase the production and population size of each of the relevant species of sea turtles.
- Description of the measures taken by Mexico and other nations to conserve each of the relevant species of sea turtles in their waters and on their shores, along with a description of the efforts to enforce these measures and an assessment of the success of these measures.
- Identification of nesting and/or reproductive locations for each of the relevant species of sea turtles in the waters and on the shores of the United States, Mexico, and other nations and of measures that

should be undertaken at each location, as well as description of worldwide efforts to protect such species of turtles.

Accordingly, a study committee was convened by the National Research Council's Board on Environmental Studies and Toxicology in collaboration with its Board on Biology. The committee included experts in international and domestic sea turtle biology and ecology, coastal zone development and management, commercial fisheries and gear technology, marine resources, and conservation biology. During the course of the committee's 1-year study, it heard from representatives of the shrimping industry, conservation organizations, the U.S. Fish and Wildlife Service, the National Marine Fisheries Service, and Sea Grant programs. The committee observed shrimp trawling exercises with and without turtle excluder devices on a converted shrimp trawler in Georgia coastal waters. It reviewed pertinent published literature and analyzed original data sets on aerial and beach turtle surveys, shrimp trawling efforts, other commercial fisheries, turtle strandings, and other materials from a variety of organizations and experienced individuals.

The present report reviews available scientific and technical information on the biology, reproductive dynamics, behavior, and distribution of five species of sea turtles. It also describes and assesses the sources of mortality incurred by the species and the effectiveness of current and required conservation measures. The committee was not charged or constituted to address and did not analyze social and economic issues related to sea turtle conservation.

2

Biology

Of the world's 12 living families and approximately 250 species of turtles, only two families, together comprising eight species, are marine. The eight species have several common characteristics, including relatively nonretractile extremities, extensively roofed skulls, and limbs converted to paddle-like flippers with one or two claws and little independent movement of the digits. All are large turtles, with adult body weights of 35-500 kg, and all show various adaptations to the marine environment, such as large salt glands to excrete the excess salt ingested with seawater and food.

The seven species of the family Cheloniidae, the hard-shelled turtles (as opposed to the leatherback in the family Dermochelyidae), have widely divergent and often specialized feeding habits: for example, the green turtle is an herbivore, and the hawksbill subsists largely on sponges. Reproductive behavior patterns are similar among the species, but some interesting variations are known. Each female lays about 100 eggs in a sand-covered cavity above the beach high-tide line, and, after an incubation period of about 2 months, hatchlings emerge usually at night. For most and probably all species, the sex of the hatchlings depends on incubation temperature.

Historically, all sea turtles have been valuable to humans. The leatherback, although widely reputed to be inedible, is killed extensively

FIGURE 2-1 Sea turtles found in U.S. coastal waters. Source: Modified from Ross et al., 1989.

for its meat and its eggs are eagerly sought as food. The green turtle is famous for "turtle soup" and steaks, and the hawksbill's "tortoiseshell" has been used for centuries in making ornamental articles. The olive ridley has been used for leather in recent decades and has served as a source of both flesh and eggs for human consumption. The loggerhead, whose shell lacks decorative appeal, is sought in some areas for its flesh and eggs.

This chapter describes five of the eight species of sea turtles (Figure 2-1) each in terms of its distribution, its population and habitats, its food habits, its reproduction and growth, and major threats to its survival. The species are discussed here in the order of their apparent need for immediate protection in U.S. waters.

KEMP'S RIDLEY

General Description

The Kemp's ridley is a small sea turtle, with adult females measuring 62-70 cm in straight carapace (upper shell) length (SCL) and weighing 35-45 kg. Adults are olive green above and yellowish below. The Kemp's ridley is slightly larger and heavier, is lighter in color, and has a lower and wider carapace than its congener, the olive ridley. Its head is large, with strongly ridged, powerful, and massive jaws. The carapace almost always has five pairs of costal scutes (scales) and usually five vertebral scutes. The hatchlings are dark gray, weigh about 17 g, and are approximately 44 mm in carapace length. The committee's interim report dealt with this species (Appendix B).

Population Distribution and Habitats

Foraging Areas

Although most Kemp's ridleys are found in the Gulf of Mexico (Hildebrand, 1982), they also occur along the Atlantic coast as far north as Long Island and Vineyard Sound, Massachusetts. Drifting hatchlings and young juveniles from the western gulf gyres apparently enter the eastern gulf loop current and are carried via the Florida current into the Gulf Stream and up the east coast (Carr, 1980; Collard, 1987). Hendrickson (1980) and Carr (1980) speculated that these young turtles were "waifs" and possibly lost to the population. The numbers returning from the northern excursion are unknown. Juvenile Kemp's ridleys tagged in the Cape Canaveral region move north with warming water and then south as water temperatures drop in the winter; that pattern suggests that the turtles have the migratory capability to move back into the Gulf of Mexico (Henwood and Ogren, 1987). Nevertheless, only when some of the many Kemp's ridleys now being tagged along the east coast are found in the gulf will their recruitment back into the breeding population be certain.

Adults are found almost entirely in the Gulf of Mexico, where tag returns from cooperative shrimp fishermen from the United States and Mexico suggest an approximately equal distribution between the northern gulf and the southern gulf (Pritchard and Márquez M., 1973; Márquez M., in prep.). Satellite-tracked females migrating north and south of the nesting beach at Rancho Nuevo remained in nearshore waters less than 50 m deep and spent less than an hour each day at the surface (Byles, 1989), an observation that reinforces the belief that the Kemp's ridley is largely a benthic species.

In the northern Gulf of Mexico, juveniles are most common between Texas and Florida (Ogren, 1989). There is no unequivocal evidence of juveniles in the southern gulf. Declining water temperatures apparently induce juveniles to move from shallower coastal areas presumably to deeper, warmer waters (pers. comm., J. Rudloe, Gulf Specimens Marine Laboratory, Panacea, Florida, 1989).

Hatchlings spend many months as surface pelagic drifters (Carr, 1980; 1986a). How long they stay in this habitat, what they eat while there, and how they get back to the coastal regions are all unknown, although Collard (1987) summarized open-water observations of the species in the gulf. The life history of the Kemp's ridley might be easier to elucidate than that of other sea turtles, because it has a more restricted distribution and nesting location in the semi-enclosed Gulf of Mexico.

Rudloe (pers. comm., Gulf Specimens Marine Lab, Panacea, Florida, 1989) suggested that during the postpelagic stages, the body size of Kemp's ridleys is positively correlated with water depth. In Louisiana, northwest Florida, and New York, the smallest juveniles are found in shallow water of bays or lagoons, often foraging in less than a meter of water (Ogren, 1989). Larger juveniles and adults probably forage in open gulf waters.

Byles (1988) radio-tracked juvenile Kemp's ridleys in Chesapeake Bay, where he reported that they used the estuary for summer feeding, but differed in habitat preference and behavior from loggerheads, which he also tracked: "The loggerheads . . . fed primarily on horseshoe crabs, *Limulus polyphemus*. The [Kemp's] ridleys, in contrast, occupied shallower foraging areas over extensive seagrass beds (*Zostera marina* and *Ruppia maritima*), did not range as far with the tide and fed mostly on blue crabs (*Callinectes sapidus*). Strong site tenacity was displayed by both species once foraging areas were established." P. Shaver and D. Plotkin (pers. comm., Padre Island National Seashore, 1989) believed that loggerheads and Kemp's ridleys partitioned food resources in Texas: the ridleys forage in shallower water and take the relatively fast blue and spotted crabs, whereas loggerheads are in deeper water and feed on slow-moving crabs.

Marine areas within several kilometers of the nesting beach at Rancho Nuevo, Mexico, constitute important internesting habitat for Kemp's ridleys. Satellite and radio-tracking studies have shown that the Kemp's ridley can wander many kilometers in Mexican waters from the nesting beach between nesting periods. Some mating occurs in March and April near the Rancho Nuevo nesting beach. Persistent reports of large numbers of Kemp's ridleys just south of the Mexico-U.S. border before the nesting season also indicate that social and mating aggregations might occur many kilometers from the nesting beach. More observations are needed regarding this poorly known aspect of Kemp's ridley biology.

Nesting Areas

The nesting beach at Rancho Nuevo is the primary terrestrial habitat for Kemp's ridleys, at about latitude 23°N on the Gulf of Mexico. Until recently it was a fairly steep sand-covered beach. During hurricane Gilbert in 1988, the beach was scoured, and that left a mixture of gravel, sand, and rock rubble. In 1989, females returned to Rancho Nuevo as in the past, but the primary nesting area was extended from the usual 15 km of beach an additional 15 km or more northward (pers. comm., J. Woody, USFWS, 1989). Only rarely has any substantial nesting been observed at any other beach (such as at Tecolutla, Veracruz; see Ross et al., 1989, for other scattered nesting sites). Nesting on the beach at Rancho Nuevo is clearly crucial to the species survival.

Food Habits

Hatchlings move quickly through the surf zone and into the pelagic zone of the Gulf of Mexico. Their feeding habits have not been observed in the wild, but it is presumed that they eat swimming and floating animal matter in the epipelagic zone.

Juveniles, subadults, and adults feed on various species of crabs and other invertebrates (Dobie et al., 1961). In the northeastern United States, where juveniles are found, crabs of several species are common in stomach contents, whereas in the Gulf of Mexico, the blue crab is the most common item. Stranded dead Kemp's ridleys often have fish parts, shrimp, and small gastropods in their guts, even though they appear to be too slow to catch these animals in the wild. Perhaps they learn to feed on the bycatch dumped overboard from trawlers (Shoop and Ruckdeschel, 1982; Manzella et al., 1988). Surprisingly, Kemp's ridleys raised in the laboratory on nonliving food will commonly capture and feed on live crabs as soon as crabs are provided. Food habits of adults in the southern Gulf of Mexico are not well documented.

Reproduction and Growth

Reproduction of the Kemp's ridley is different from that of other U.S. sea turtle species in four important ways. First, it nests in an aggregated fashion; many females gather in the sea near the nesting beach and then emerge to nest in a loosely synchronized manner over several hours in what is known as an "arribada" or "arribazon" pattern. An important amateur movie made by Andrés Herrera in 1947 documented an arribada of approximately 40,000 females nesting on one day at the Rancho Nuevo beach (Carr, 1963; Hildebrand, 1963). Second, Kemp's ridleys nest during the daytime, whereas the other species nest at night. Solitary nesters, arribada groups, and even most captive reared females nest exclusively during the daytime. Third (and unique for sea turtles), almost all nesting occurs at one site—a site in the state of Tamaulipas, Mexico, near Rancho Nuevo. Exceptions are occasional nests in Texas, a single recent nest in Florida, and a potentially important but irregular nesting area near Tecolutla, Veracruz. Fourth, most females nest annually.

The most reliable index of Kemp's ridley population size has been the annual count of nesting adult females at Rancho Nuevo. The number of nesting females there decreased from an estimated 40,000 (in a single day) in 1947 to an estimate of about 650 throughout the nesting season in 1988; the latter number was based on the total of 842 nests found (Ross et al., 1989; Appendix B). In 1989, even including the newly found extension of the nesting beach some 15 km to the north, the total number of nests found (784) signaled a further decline of this species (pers. comm., J. Woody, USFWS, 1989).

As with other sea turtles, both males and females migrate toward the nesting area, and courtship and mating probably occur during several weeks before the female emerges to nest (Owens, 1980). Studies of captive animals indicate that a single mating receptivity period is regulated by the female and occurs about 4 weeks before the first nest is dug (Rostal et al., 1988). After the mating, fertilized eggs are stored in the oviduct until nesting. Nesting is usually restricted to April, May, and June—and occasionally July, if a cool spring delays the onset of reproduction.

A female deposits one to four clutches per season, laying an average of about 105 eggs per clutch (Márquez M., in prep.). Most females nest annually, based on the return of tagged animals, but they can also skip a year. Because nesting occurs over only about 45 minutes and because many turtles might nest simultaneously over several kilometers of beach, it has been difficult to tag or check for previous tags on every nesting female. Data on the nesting biology of the Kemp's ridley are, therefore, still incomplete. With the limited data available, the number of adult females in the world population can be estimated from the equation:

$$P_{nf} = \frac{N_t}{N_f} \div \rho_{nf}$$

where

P_{nf} = total population of adult females
N_t = total number of nests per year
N_f = average number of nests per reproductively active female
ρ_{nf} = proportion of females that nest in a given year

Observers who have worked closely with Kemp's ridleys argue that the actual number of nests per year per female is not 1.3, as suggested by Márquez M. et al. (1981), but may be about 2.3 per year (Pritchard, 1990). If this proves to be true, nesting females are far fewer than was previously thought—about 350, rather than 620 per year on average from 1978 to 1988. A firm estimate of P_{nf} is still not available.

The incubation time of Kemp's ridley eggs averages 50-55 days (Ross et al., 1989). Growth rates of wild hatchlings are unknown, and the smallest wild juveniles (about 20 cm SCL) found in the northern Gulf of Mexico are of unknown age. In captivity, on a carefully prepared high-protein diet, they can grow to 20 cm in 10-18 months (Klima and McVey, 1982). However, it might take 2 years or longer for Kemp's ridleys to reach that size in the pelagic zone. Standora et al. (1989) tracked and recaptured three juvenile Kemp's ridleys in Long Island Sound. They averaged about 6 kg in weight and gained 548 g/month during the summer. Animals hatched in captivity, released, and then recaptured after 2 years or more grew at rates that suggested that the turtles could reach adult size in 6 or 7 years. Márquez M. (in prep.) noted that many females continue to grow slowly after reaching maturity.

Age, size structure, and sex ratios of the population are poorly known. Recent stranding records and the work of Ogren (1989) and collaborators suggest an increase in recruitment of small juveniles into the coastal habitats of the species in recent years. Danton and Prescott (1988) found a male-to-female ratio of 20:28 in 48 stranded dead juvenile Kemp's ridleys (mean SCL = 27.1 cm) from Cape Cod, Massachusetts. Although the sample size is small, it does indicate that the ratio is not strongly skewed. Sex ratios for other sizes and places have not been determined for Kemp's ridleys. Sex of a developing Kemp's ridley is dependent on the temperature of egg incubation, on the basis of work with headstarted animals (Shaver et al., 1988). At higher incubation temperatures, more females are produced.

Major Threats to Survival

At various stages of their life cycle, Kemp's ridleys can be adversely affected by a number of activities and substances. These potentially include cold-stunning; human and nonhuman predation of eggs in nests; predation of hatchlings and/or older turtles by crabs, birds, fish, and mammals, including humans from foreign nations; ingestion of plastics; industrial pollutants; diseases; exploratory oil and gas drilling; dredging; explosive removal of oil platforms; and incidental capture in shrimping and other fishing gear. The relative impacts of these mortality factors are discussed in Chapter 6.

LOGGERHEAD

General Description

Adult and subadult loggerheads have reddish-brown carapaces and dull brown to yellowish plastrons (lower shells). The thick, bony carapace is covered by nonimbricate horny scutes, including five pairs of costals, 11 or 12 pairs of marginals, and five vertebrals. Adult loggerheads in the southeastern United States have a mean SCL of about 92 cm and a mean body weight of about 113 kg, but adults elsewhere are usually smaller (Tongaland, Hughes, 1975; Colombia, Kaufmann, 1975; Greece, Margaritoulis, 1982). They rarely exceed 122 cm SCL and 227 kg. The brown hatchlings weigh about 20 g and are 45 mm long.

Population Distribution and Habitats

Foraging Areas

The geographic distribution of loggerheads includes the subtropical (and occasionally tropical) waters and continental shelves and estuaries along the margins of the Atlantic, Pacific, and Indian Oceans. It is rare or absent far from mainland shores. In the Western Hemisphere, it ranges as far north as Newfoundland (Squires, 1954) and as far south as Argentina (Frazier, 1984) and Chile (Frazier and Salas, 1982).

Nesting Areas

Nesting is concentrated in the north and south temperate zones and subtropics with a general avoidance of tropical beaches in Central America, northern South America, and the Old World. The largest known nesting aggregation was reported on Masirah and the Kuria Muria Islands of Oman (Ross and Barwani, 1982), and a nesting assemblage has been noted recently on the Caribbean coast of Quintana Roo (pers. comm., R. Gil, Quintana Roo, Mexico, 1989). In the western Atlantic, most nesting occurs on Florida beaches, with approximately 90% in Brevard, Indian River, St. Lucie, Martin, Palm Beach, and Broward counties. Nesting also occurs regularly in Georgia, South and North Carolina, and along the gulf coast of Florida.

Aerial beach surveys in 1983 estimated that 58,016 nests were dug along the southeastern United States (Murphy and Hopkins, 1984) and provided the best estimate of population size. Assuming a mean of 4.1 nests per female, approximately 14,150 females nested on the southeast coast in 1983 (Murphy and Hopkins, 1984). Those nests constitute about 30% of the known worldwide nesting by loggerheads and clearly rank the southeastern U.S. aggregation as the second largest in the world, only the Oman assemblage being larger (Ross, 1982).

Recently, Witherington and Ehrhart (1989a) concluded that the stock of loggerheads represented by adult females that nest in the southeastern U.S. is declining. Evidence of a decline came from the current best estimates of adult females nesting each year (Murphy and Hopkins, 1984), published life tables and population models (Richardson and Richardson, 1982; Frazer, 1983b; Crouse et al., 1987), observed mortality rates in the southeastern United States, and observed population declines in South Carolina (pers. comm., S. Murphy, S.C. Wildlife and Marine Resources, 1989) and Georgia (pers. comm., J.I. Richardson, University of Georgia, 1989).

Adult females generally select high-energy beaches on barrier strands adjacent to continental land masses for nesting. Steeply sloped beaches with gradually sloped offshore approaches are favored (Provancha and Ehrhart, 1987). After hatching and leaving the beach, hatchlings apparently swim directly offshore and eventually associate with sargassum and debris in pelagic drift lines that result from current convergences (Carr, 1986a; 1987). The evidence suggests that posthatchlings that become a part of the sargassum raft community remain there as juveniles, ride current gyres for possibly several years, and grow to 40-50 cm SCL. They then abandon the pelagic habitat, moving into the nearshore and estuarine waters along continental margins, and use those areas as the developmental habitat for the subadult stage. In such places as the Indian River Lagoon, Florida, the subadults are separated from the adults, whose foraging areas are apparently hundreds of kilometers away. Nothing is known about the transition from subadult to adult foraging areas, but it seems clear that adults can use a variety of habitats including the Atlantic continental shelf. Remote recoveries of females tagged in Florida indicate that many migrate to the Gulf of Mexico, often to the turbid, detritus-laden, muddy-bottom bays and bayous of the northern gulf coast (Meylan et al., 1983). Others apparently occupy the clear waters of the Bahamas and Antilles, with sandy bottoms, reefs, and shoals that constitute a totally different type of habitat. Nothing is known of the periods of time that loggerheads spend in these disparate habitats or of their propensity to move from one to another.

Food Habits

Although the list of food items used by loggerheads is long and includes invertebrates from eight phyla (Dodd, 1988), subadult and adult loggerheads are primarily predators of benthic mollusks and crustaceans. Coelenterates and cephalopod mollusks are especially favored by loggerheads in the pelagic stage (van Nierop and den Hartog, 1984). Posthatchling loggerheads evidently ingest macroplankton associated with "weed lines," especially gastropods in the sargassum raft community as well as fragments of crustaceans and sargassum (Carr and Meylan, 1980). Loggerheads sometimes scavenge fish or fish parts or incidentally ingest fish (Brongersma, 1972).

Reproduction and Growth

It has been assumed for some time that, at least for Florida loggerheads, males migrate with females from distant foraging areas to the waters off nesting beaches, where courtship and mating take place. Mating takes place in late March to early June (Caldwell, 1959; Caldwell et al., 1959a; Fritts et al., 1983). Although a few adult males might remain off the Florida coast throughout the year (Henwood, 1987), most of them apparently depart by about mid-June. Females mate before the nesting season during a single receptive period and then lay multiple clutches in nests dug in the beaches throughout some portion of the nesting season

(Caldwell et al., 1959b). Mean clutch size varies from about 100 to 126 along the southeastern United States coast.

In the southeastern United States adult females begin to nest as early as the last week of April; nesting reaches a peak in June and July and continues until early September. Loggerheads nest one to seven times per season (Talbert et al., 1980; Lenarz et al., 1981; Richardson and Richardson, 1982); the mean is believed to be approximately 4.1 (Murphy and Hopkins, 1984). The internesting interval is about 14 days.

Loggerheads are nocturnal nesters, with infrequent exceptions (Fritts and Hoffman, 1982; Witherington, 1986). Good descriptive accounts of loggerhead nesting behavior have been given by Carr (1952), Litwin (1978), and Caldwell et al. (1959a). Remigration intervals of two and three years are most common in loggerheads, but the number can vary from one to six years (Richardson et al., 1978; Bjorndal et al., 1983).

Natural incubation periods for United States loggerheads are about 54 days in Florida (Davis and Whiting, 1977; Witherington, 1986), about 63 days in Georgia (Kraemer, 1979), and about 61 days in North Carolina (Ferris, 1986). The length of the incubation period is inversely related to nest temperature (McGehee, 1979), and the sex of loggerhead hatchlings also depends on temperature (Yntema and Mrosovsky, 1980; 1982). Hatching success has been reported at 73% and 55% in South Carolina (Caldwell, 1959) and 56% in Florida (Witherington, 1986).

Growth rates of captive posthatchling and juvenile loggerheads have been reported (e.g., Witham and Futch, 1977), but no data are available on these stages in the wild. In captivity, young loggerheads can grow to about 63 cm SCL and 37 kg in 4.5 years (Parker, 1926). In wild subadults, linear growth rates vary from 1.5 cm/year in Australia (Limpus, 1979) to 5.9 cm/year in Florida (Mendonca, 1981). Growth rates of larger subadults decrease with increasing carapace length. Frazer and Ehrhart (1985) estimated age at maturity as 12-30 years.

Hatchlings engage in a "swimming frenzy" for about 20 hours after they enter the sea, and that frenzy takes them 22-28 km offshore (Salmon and Wyneken, 1987). They become associated with sargassum rafts or debris at current rips and other surface water convergences and begin the juvenile life stage (Carr, 1986b). After perhaps 3-5 years circumnavigating the Atlantic in current gyres (Carr, 1986a) or after reaching 45 cm SCL, they abandon the pelagic environment and migrate to nearshore and estuarine waters along the eastern United States, the Gulf of Mexico, and the Bahamas to begin their subadult stage. Henwood (1987) reported a tendency for subadults of the Port Canaveral aggregation to disperse more widely in the spring and early summer. Chesapeake Bay subadults exhibit a variety of movements between waters of different temperature and salinity (Killingly and Lutcavage, 1983). Recoveries of females tagged while nesting on the Florida east coast suggest that they dispersed widely to foraging areas in the Gulf of Mexico, in Cuba, elsewhere in the Greater Antilles, and in the Bahamas (Meylan et al., 1983). Those females apparently remigrate hundreds of kilometers at multiyear intervals to nest on the preferred, high-energy nesting beaches of eastern Florida. Much less is known about migrations of Georgia, South Carolina, and North Carolina nesters outside the nesting season, because of the dearth of reported tag recoveries. Females from Georgia dispersed along the Atlantic seaboard and did not appear in tropical waters outside the United States (Bell and Richardson, 1978).

Major Threats to Survival

Loggerheads are subject to numerous threats to their survival, including egg-collecting, raccoon predation on nests and eggs, and a variety of human activities such as beachfront development, increases in artificial illumination and disturbance, and incidental capture in shrimping and other fishing gear. They are also subject to effects of oil-platform removal, dredging, ingestion of plastics, and boat collisions. The relative impacts of these mortality factors are discussed in Chapter 6.

GREEN TURTLE

General Description

The green turtle is the largest hard-shelled sea turtle. Adults have a carapace varying in color from black to gray to greenish or brown, often with bold streaks or spots, and a yellowish white plastron. Populations around the world differ greatly in adult size and weight; those in Florida average 101.5 cm SCL and 136.2 kg body weight (Witherington and Ehrhart, 1989a). Characteristics that distinguish them from other sea turtles are their small, rounded head, smooth carapace, and four pairs of costal scutes. Hatchlings weigh approximately 25 g, their black carapace is about 50 mm long, and the ventral surface is white.

Population Distribution and Habitats

Foraging Areas

The circumglobal distribution in tropical and subtropical waters has been described by Groombridge (1982). In U.S. Atlantic waters, green turtles occur around the U.S. Virgin Islands and Puerto Rico and from Texas to Massachusetts. Important feeding areas for green turtles in Florida include the Indian River, Florida Bay, Homossassa Bay, Crystal River, and Cedar Key. Those areas and the Texas coast (Aransas Bay, Matagorda Bay, and Laguna Madre) figured heavily in the commercial fishery for green turtles at the end of the last century (Hildebrand, 1982; Doughty, 1984).

Green turtles occupy three habitat types: high-energy beaches, convergence zones in the pelagic habitat, and benthic feeding grounds in relatively shallow, protected waters. Hatchlings leave the beach and apparently move into convergence zones in the open ocean (Carr, 1986a). When they reach 20-25 cm SCL, they leave the pelagic habitat and enter benthic feeding grounds. The foraging habitats are most commonly pastures of seagrasses or algae, but small green turtles are also found over coral reefs, worm reefs, and rocky bottoms. Some feeding grounds support only particular size classes of green turtles; the turtles apparently move among these developmental feeding grounds. Other feeding areas, such as Miskito Cays, Nicaragua, support a complete size range of green turtles from 20 cm to breeding adults. Coral reefs and rocky outcrops near feeding pastures often are used as resting areas.

The navigation feats of the green turtle are well known, but poorly understood. Hatchlings and adult females on the nesting beach use photic cues to orient toward the ocean (Ehrenfeld, 1968; Mrosovsky and Kingsmill, 1985). Unknown are the cues used in pelagic-stage movements, in movements among foraging grounds, or in migrations between foraging grounds and the nesting beach. Because green turtles feed in marine pastures in quiet, low-energy areas and nest on high-energy beaches, their feeding and nesting habitats are, of necessity, some distance apart. Green turtles that nest on Ascension Island forage along the coast of Brazil, well over 1,000 km away (Carr, 1975). The location of the foraging grounds of green turtles that nest in Florida is not known, and individuals foraging in Florida waters might not be part of the nesting population there. It has been generally accepted, but not proved, that green turtles return to nest on their natal beach. Green turtles do exhibit strong site fidelity in successive nesting seasons. Meylan (1982) has reviewed information on turtle movements based on tag returns.

Nesting Areas

Females deposit egg clutches on high-energy beaches, usually on islands, where a deep nest cavity is dug above the highwater line. Major green turtle nesting activity occurs on Ascension Island, Aves Island, in Costa Rica, and in Surinam. In U.S. Atlantic waters, green turtles nest in small numbers in the U.S. Virgin Islands and in Puerto Rico and in somewhat larger numbers in Florida, particularly in Brevard, Indian River, St. Lucie, Martin, Palm Beach, and Broward counties.

Food Habits

Posthatchling, pelagic-stage green turtles are presumably omnivorous, but dietary data are lacking. When green turtles shift to benthic feeding grounds, they prefer to feed on seagrasses and macroalgae. Details of diet and nutrition of green turtles have been reviewed by Mortimer (1982a) and Bjorndal (1985).

Reproduction and Growth

Green turtles mate in the water off the nesting beaches. Evidence is accumulating that males might migrate to the nesting beach every year (Balazs, 1983). Females emerge at night to deposit eggs; the nesting process takes about 2 hours. Descriptions of their behavior have been reviewed by Ehrhart (1982). The females deposit one to seven clutches in a breeding season at intervals of 12-14 days. The average number of clutches is usually stated as two to three (Carr et al., 1978), but might be more. Mean clutch size is usually 110-115 eggs, but it varies among populations. The average egg count reported for 130 Florida clutches was 136 (Witherington and Ehrhart, 1989a). Only occasionally do females produce clutches in successive years; usually 2 years or more pass between breeding seasons.

Hatching success of undisturbed nests is usually high, but predators destroy a high percentage of nests on some beaches (Stancyk, 1982). Many nests are also destroyed by tidal inundation and erosion. As with some other species, hatchling sex depends on incubation temperature (Standora and Spotila, 1985). Hirth (1980), Ehrhart (1982), and Bjorndal and Carr (1989) have reviewed the reproductive biology of green turtles.

The numbers of recorded nestings in Florida were 736 in 1985, 350 in 1986, 866 in 1987, and 446 in 1988 (Conley and Hoffman, 1987; unpublished data, Florida Department of Natural Resources). It is impossible to assess trends in the nesting population from these data because the length of beach surveyed varied among years: 616 km in 1986, 832 km in 1987, and 971 km in 1988 (unpublished data, Florida Department of Natural Resources, 1988).

Green turtles grow slowly. Rates of pelagic-stage green turtles have not been measured under natural conditions, but growth rates have been measured on the benthic feeding grounds. In the southern Bahamas, they grew from an SCL of 30 cm to an SCL of 75 cm in 17 years, and linear growth rate decreased with increasing carapace length (Bjorndal and Bolten, 1988). Estimates of age at sexual maturity range from 20 to 50 years (Balazs, 1982; Frazer and Ehrhart, 1985).

Major Threats to Survival

Over much of its range, the green turtle has been severely depleted because of high demand for both eggs and meat as human food. Exploitation has been intense on both nesting beaches and foraging grounds, and cannot be reversed quickly, because the green turtle takes several decades to reach maturity. Degradation of nesting and feeding habitats are also serious problems.

HAWKSBILL

General Description

Adult hawksbills are easily recognized by their thick carapace scutes, often with radiating streaks of brown and black on an amber background, and a strongly serrated posterior margin of the carapace. Their common name is derived from the narrow head and tapering "beak." Except for Kemp's ridley, the hawksbill is the smallest of the five species, with an SCL less than 95 cm. A sample of 121 nesting females from several localities around the Caribbean averaged 81 cm SCL (range, 62.5-91.4 cm) (Witzell, 1983). Hatchlings are brown to nearly black.

Population Distribution and Habitats

Foraging Areas

Hawksbills typically forage near rock or reef habitats in clear shallow tropical waters (Witzell, 1983). That habitat is preferred for feeding on encrusting organisms, particularly some sponges. Hawksbills observed off the shore of Antigua (pers. comm., J. Fuller, Antigua, 1989) and Mona Island, Puerto Rico (Kontos, 1985) appear to be associated with benthic feeding territories, with the deeper territories used by the larger animals. Hawksbills associate with a variety of reef structural types from vertical underwater cliffs to gorgonian flats. Adults usually are not found in shallow marine habitats (less than 20 m deep) near land, whereas small juveniles are never far from the shallowest coral reefs. Much of the Caribbean down to 100 m or even more might provide foraging habitat for adults, because sponges grow well to these depths.

Hawksbills are found throughout the Caribbean and are commonly observed in the Florida Keys, in the Bahamas, and in the southwestern Gulf of Mexico. They are not reported as frequently from shallow coastal systems with soft bottoms and high turbidity, such as the eastern U.S. coast north of Cape Canaveral. However, small juvenile hawksbills have been caught in shallow nearshore areas of the Guianas, characterized by very muddy water, and adults have nested on adjacent beaches.

Offshore behavior of hawksbills is poorly understood. Adults (singly or in mated pairs) and large juveniles are commonly seen in all seasons well off the shore of Antigua and Barbuda in water up to 100 m deep (pers. comm., J. Fuller, Antigua, 1989). Presumably, these animals are foraging, their presence suggesting an ability to dive to considerable depths to feed on live bottom-sponges.

During the pelagic phase, hatchlings presumably associate with sargassum rafts in the Caribbean. Young individuals first appear as foraging residents of shallow reef systems when they reach 15-25 cm SCL. Hawksbills might be much more sedentary than other members of the family Cheloniidae (Witzell, 1983), but long-range tag returns indicate that hawksbills can move hundreds of kilometers between their nesting beaches and foraging areas (Nietschmann, 1981; Parmenter, 1983; Bjorndal et al., 1985). When a young hawksbill changes from a pelagic feeder to a benthic-reef feeder, it apparently uses a foraging territory that it stays in until it shifts its foraging territory, probably moving from shallow to deep water as it becomes capable of deeper dives. Whether a neophyte breeder returns to the proximity of its natal origin is unknown.

Understanding of neonate movements at sea is speculative. Prevailing winds and currents would carry Antillean hatchlings into a Caribbean sargassum gyre, with some transportation possible on currents north along the Yucatan coast to the western Gulf of Mexico. There is no evidence that Caribbean hawksbill hatchlings use the North Atlantic Sargasso Sea and its associated gyre, as U.S. Atlantic loggerheads apparently do. In the Azores, where many young loggerheads are found, juvenile hawksbills are not known.

Nesting Areas

Hawksbills nest on tropical islands and sparsely inhabited tropical continental shores around the world. Eastern Atlantic nesting records are from only a few African locations and associated offshore islands (Brongersma, 1982). Western Atlantic nesting records extend from Brazil to Florida's southern Atlantic coast and include the islands and continental coastline of the Caribbean and the southwestern Gulf of Mexico (Campeche). Substantial nesting might occur on the continent and the offshore keys around the Caribbean and Lesser Antilles (Witzell, 1983; Pritchard and Trebbau, 1984). Although the hawksbill is often described as a dispersed nester (Pritchard and Trebbau, 1984), small nesting concentrations do exist on Antigua, for example. However, nesting generally is distributed at low densities across much of the Caribbean.

Nesting within U.S. waters follows the same pattern as in the Caribbean at large. Scattered nesting can occur on almost any beach of the U.S. Virgin Islands (Boulon, 1983), Puerto Rico (including Vieques (Pritchard and Stubbs, 1982) and the Culebra group (Meylan, 1989), or southern Florida (Lund, 1985; McMurtray and Richardson, 1985). Higher nesting concentrations are found on remote islands, such as Mona Island off Puerto Rico (Thurston and Wiewandt, 1975; Olson, 1985; Kontos, 1988; Tambiah, 1989) and Buck Island in the Virgin Islands (Hillis and Mackay, 1989a).

Nesting habitat varies from high energy ocean beaches shared with green turtles (Carr and Stancyk, 1975) to tiny pocket beaches several meters wide contained in the crevices of cliff walls. A typical nesting habitat is a low-energy sand beach with woody vegetation, such as seagrape or saltshrub near the water line. Some active nesting beaches have no exposed sand, but have woody vegetation growing to the water's edge. In contrast, hawksbills at Sandy Point, St. Croix, regularly traverse 30 m of open sand to reach an acceptable nesting habitat (pers. comm., K. Eckert, University of Georgia, 1989). A portion of the nesting beach in Antigua with vegetation set 30 m back from the water's edge is rarely used (pers. comm., J. Richardson, University of Georgia, 1989), but turtles nest regularly on either side where the vegetation is closer to the water.

Food Habits

Until recently, hawksbills were considered to be generalists, feeding on a wide variety of marine invertebrates and algae (Carr and Stancyk, 1975; Witzell, 1983). But Meylan (1988) showed that hawksbills specialize on sponges, selecting just a few genera throughout the Caribbean. Much of the other material in hawksbill stomachs was apparently ingested coincidentally while the animals were feeding on sponges. Neonates in captivity appear to do well on a diet of sargassum (Pritchard and Trebbau, 1984).

Reproduction and Growth

The predominant nesting months for hawksbills in Puerto Rico and the U.S. Virgin Islands are June to November, although some nesting can be documented for every month of the year (Witzell, 1983). Adult females can make their first appearance at a nesting beach any time from June to September. If a population contains only a few animals, females that use a particular nesting beach might arrive rather irregularly, causing the apparent nesting season to vary widely from year to year. Such events might explain the differences in nesting seasons observed on Buck Island in the Virgin Islands over the last 10 years (Hillis and Mackay, 1989a).

The modal number of nests per female during a single season in Antigua is five; individuals nest four to six times (Corliss et al., 1989). Estimates of clutches per year (Witzell, 1983) less than the Antigua number possibly result from inadequate beach coverage, as has been documented for other sea turtles (Tucker, 1989a).

The interval between consecutive clutches averages 14 days in Antigua (Corliss et al., 1989) and Mona Island (Kontos, 1988), 16 days at Tortuguero, Costa Rica (Bjorndal et al., 1985), and 18.5 days in Nicaragua (Witzell, 1983).

The modal remigration interval of nesting hawksbills is 3 years at Tortuguero, Costa Rica (Carr and Stancyk, 1975). An intensive survey of nesting hawksbills in Antigua produced no records of annual remigration, but 17 of 23 nesting turtles in 1989 had been tagged at the same beach in 1987 (Corliss et al., 1989). These preliminary results suggest a dominant 2-year remigration interval.

Hawksbill nesting behavior has been well documented (Witzell, 1983; Pritchard and Trebbau, 1984). Individuals usually take one or more hours to complete the sequence. Clutch size varies greatly from site to site (Witzell, 1983), but the average for eastern Caribbean animals is close to 150 eggs (Corliss et al., 1989); one clutch of 215 eggs was recorded. The mean incubation time to emergence of hatchlings in Antigua was 61 days in 1987 and 68 days in 1988, with a range of 20 days around the mean (Corliss et al., 1989). Hatching success measured for several beaches averaged close to 80% (Witzell, 1983; Corliss et al., 1989).

Temperature-modulated sex ratios have not been documented in hawksbills, but are assumed to exist as in other sea turtles.

Pritchard and Trebbau (1984) reviewed information on the growth rates of captive hawksbills. Hatchlings in captivity with saturation feeding reached a carapace length of about 20 cm SCL in 1 year and 35 cm SCL in 2 years. Hatchlings in captivity can reach 50 cm SCL in 4 to 5 years. Age to maturity is not known and has not been calculated for hawksbills.

Little is known about hawksbill reproduction in the continental United States because the observed number of nests each year in Florida could have been made by as few as 1-5 females.

Major Threats to Survival

The hawksbill is considered endangered throughout its world range primarily because of widespread harvest of turtles for the international trade in tortoiseshell products, polished shells, and stuffed turtles. Killing, specimens of almost any size for their valuable scutes is widespread. Additional killing of juvenile hawksbills for trade in stuffed specimens raises mortality to catastrophic levels. The diffuse nesting habits of the hawksbill make systematic exploitation of the nesting females difficult, but also makes them hard to protect. Even when a nesting turtle escapes to the sea, the eggs commonly are taken by humans.

In addition, the hawksbill is edible and is even the preferred turtle species in a few areas. In some parts of its range, especially in the Indian Ocean, an occasional hawksbill is highly poisonous.

LEATHERBACK

General Description

The leatherback is the largest of all living sea turtles, attaining a length of 150-170 cm SCL and a weight that occasionally reaches 500 kg (rarely 900 kg). Its shell is unique in being covered with a continuous layer of thin, black, often white-spotted skin, instead of keratinized scutes. The carapace is raised into a series of seven longitudinal ridges. Other distinctive features are the absence of claws, the absence of scales (except in hatchlings and very young animals), the long forelimbs (1 m), and the reduced skeleton. Many bones that are present in the shells of other turtles are absent in the leatherback (Pritchard, 1979).

Population Distribution and Habitats

Foraging Areas

The leatherback is sometimes seen in coastal waters, but is essentially pelagic and dives to great depths. It is frequently encountered outside the tropics, even in latitudes approaching polar waters. For example, it is often reported in the waters of New England and the Maritime Provinces of Canada, possibly as far north as Baffin Island. In the southern hemisphere, records exist from Tasmania and the southern tip of New Zealand.

Nesting Areas

Leatherbacks nest almost entirely in the tropics, with extra-tropical nesting essentially confined to low-density nesting (about 20-30 turtles each year) in Florida and in South Africa. Nesting is usually colonial. The largest colonies use continental, rather than insular, beaches. In the western Caribbean, nesting is frequent from northern Costa Rica to Colombia and in eastern French Guiana and western Surinam. Some nesting also occurs along the central Brazilian coast, and important colonies are found in northwestern Guyana and in Trinidad. In the Antilles, most nesting occurs in the Dominican Republic and on islands close to Puerto Rico, including Culebra and St. Croix (U.S. Virgin Islands). The St. Croix population is the largest, best-studied one in the United States. A few nests are recorded each year on many of the islands of the Caribbean.

Leatherback nesting beaches have some common characteristics. The absence of a fringing reef appears to be important; most beaches have high-energy wave action and a steep ascent. They also have deep, rock-free sand and are adjacent to deep oceanic water. In the Guianas, adjacent waters are relatively shallow, but the presence of abundant mud and the absence of rocks or coral apparently make these beaches acceptable for nesting.

Food Habits

Leatherbacks are primarily water-column feeders, rather than benthic feeders. Many species of coelenterates, especially jellyfish, have been found in their stomachs. They have numerous adaptations of the head and mouth for their diet. Their jaws are sharp-edged and scissor-like in action, and their throat musculature is highly developed to generate a powerful inflow of water as the prey is taken. In addition, the esophagus, which might be nearly 2 m long, is lined with thousands of sharp flexible spines, which are also found in other sea turtles. Because the spines are directed toward the stomach, when the water taken in with prey is expelled, the spines retain the food.

Reproduction and Growth

Leatherbacks can travel great distances between feeding and nesting areas, and migrations of tagged animals from nesting grounds in the southern Caribbean or the Guianas to the waters of New York or New England have been recorded. One postnesting female moved from the Guianas to West Africa within a few months. However, such demanding migrations do not appear to be undertaken annually, and almost all recorded remigrations of leatherbacks to their nesting grounds have been 2 or 3 years after initial tagging. Up to 10 nestings per season per female have been recorded, with a typical leatherback internesting interval of 10 days.

Leatherback eggs are large, about 6 cm in diameter, but are not as numerous as those of other sea turtles. In the Atlantic, a typical nest includes 80-90 normal eggs but in the eastern Pacific, usually fewer than 60. Nests contain different numbers of yolkless, undersized eggs.

Eggs hatch after about 65 days. Hatching success can approach 100% in an undisturbed natural nest, but on many beaches many eggs are lost to erosion—a result of the high energy of the beaches favored by leatherbacks and the limited ability of such heavy and cumbersome animals to travel far inland to deposit their eggs. Eggs can be transferred to hatcheries, but they need even more careful handling than those of other sea turtles, if viability is to be maintained during the transfer.

Major Threats to Survival

The products of the leatherback rarely, if ever, are featured in international commerce. The common belief that this species is inedible is unfounded; intense slaughter of nesting females occurs in many areas, such as Guyana, Trinidad, Colombia, and the Pacific coast of Mexico. Even in areas where the adults are rarely killed, egg collecting might be intense. Ingestion of plastics could be an important mortality factor.

OLIVE RIDLEY

The olive ridley (*Lepidochelys olivacea*), although probably the most numerous sea turtle worldwide, is very rare in U.S. waters, and its status and future are not in the main, a direct United States responsibility. Details of its biology and reproduction can be found in Pritchard (1979).

3

Population Trends

The status of sea turtle species is perhaps best indicated by long-term changes or trends in the sizes of individual populations. Because females repeatedly return to the same beaches to nest and because this is the time in their life cycle at which they are most available for direct counting by humans, counts of nesting females or nests provide the best available long-term data on the status of their populations.

The number of nests is an index that can be correlated with population size of mature females, rather than a direct estimate, because sea turtles do not necessarily nest every year and because a female usually nests several times in a nesting season. But use of the index requires the fewest questionable assumptions about the biology of individual species of sea turtles. Other measures of long-term change have been made, such as counts from oceanic aerial surveys (Appendix D), counts of carcass strandings (Appendix E), catch per unit of effort in fishing gear, and tortoiseshell shipments to foreign markets. Some of these (aerial surveys, carcass strandings, catch in fishing gear) do not differentiate individual populations. Others depend on local changes in proximate mortality factors (carcass strandings), depend on market conditions (tortoiseshell shipments), or are expensive and have poor repeatability (aerial surveys). For those reasons, this chapter presents either trends in the number of nests or trends in the number of females tagged on a nesting beach during a season.

Short-term changes in numbers of nesting females or nests should not be interpreted as a population trend. For example, the variation in numbers of nesting green turtles at Tortuguero in the late 1970s (Figure 3-1*j*) is unrelated to absolute changes in the population size of these long-lived animals. Female green turtles return to nest every 2 or 3 years, as reflected in the year-to-year variation. Most female Kemp's ridleys apparently nest annually. Thus, the interyear variation in number of nesting females is expected to be less in Kemp's ridleys than in other species (Figure 3-1*a*).

Wide year-to-year fluctuations in numbers of nesting turtles make conclusions from short-term data sets misleading. A decade or more, depending on longevity, might be required to measure a real change in a population. For example, the Little Cumberland Island, Georgia data on loggerheads (Figure 3-1*j*) provide the results of 26 years of intensive and precisely replicated estimates at the same study site. A 10-year survey from 1964 to 1973 would have indicated no change in the population over the decade. Likewise, a 12-year survey, initiated say, in 1973 and concluded in 1984 would have produced a similar result, even though a substantial decrease in nesting females apparently occurred in the early 1970s. However, a survey from 1982 to 1989 would have suggested a progressive decline of about 10% per year. Over the entire 26-year period, 1964-1989, an average decline of about 3% per year occurred in nesting loggerheads. Thus, analyses of population trends can suggest different results depending on the years surveyed.

Surveys of a decade or less may be insufficient to indicate a population trend, as indicated above. Surveys longer than a decade become increasingly valuable for management purposes, because they can transcend short-term fluctuations that obscure long-term trends. Consequently, the results of surveys of sea turtle species provided in Figure 3-1 must be interpreted cautiously. Numbers of nests and nesting females are assumed to generate comparable and useful data on all sea turtle species.

KEMP'S RIDLEY

In 1947, an estimated 40,000 female Kemp's ridleys were observed nesting during a single day at Rancho Nuevo (Carr, 1963; Hildebrand, 1963), as judged from a motion picture taken by an amateur photographer. Data on the status of the nesting colony over the next 18 years are lacking. By 1966, the nesting assemblages or "arribadas" (aggregations of nesting females at a given place on a given day or series of days) were

FIGURE 3-1 Trends in sea turtle populations by number of nests per year (N) or number of nesting females per year (♀). D indicates isolated (nonconsecutive) years of data. (The committee has provided the following values produced by linear regression analysis: r^2, slope, and two-tailed p values.)

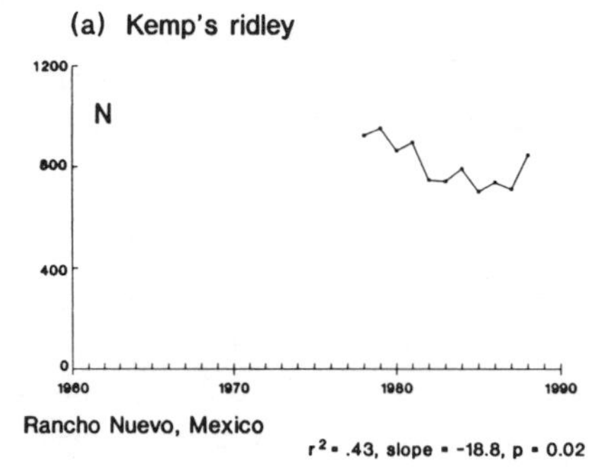

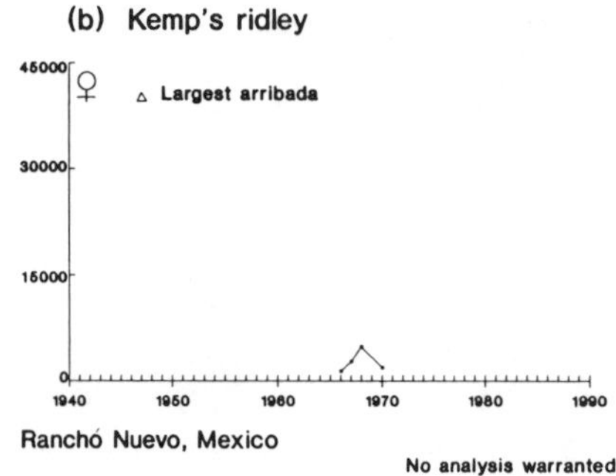

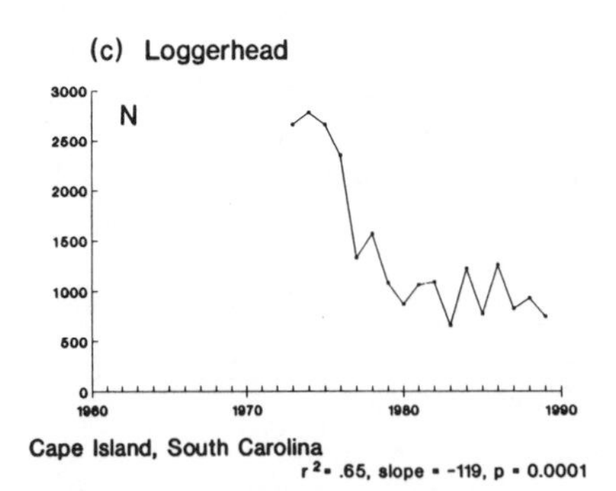

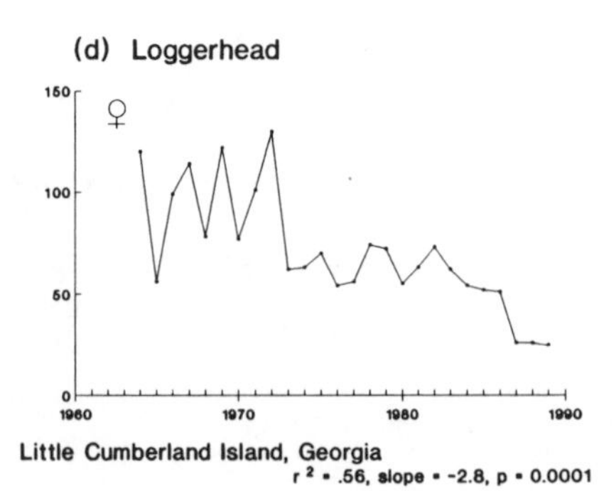

FIGURE 3-1
(Continued)

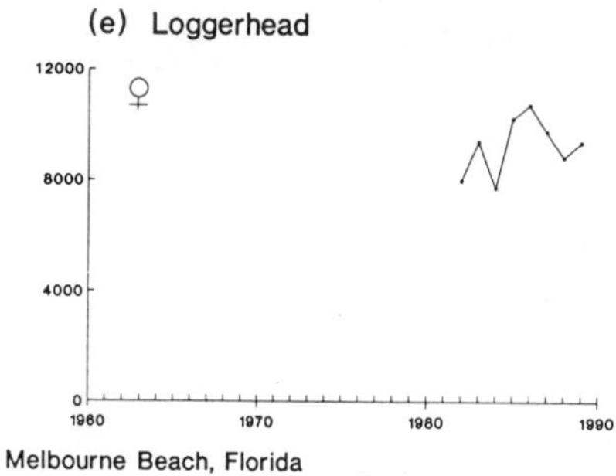

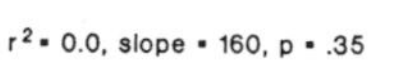

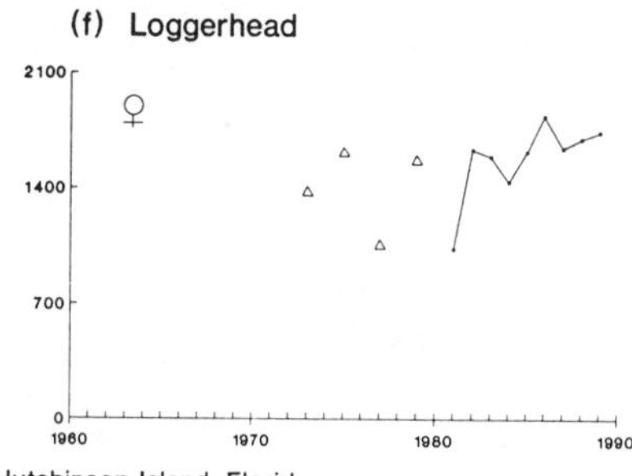

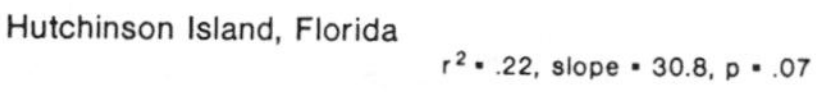

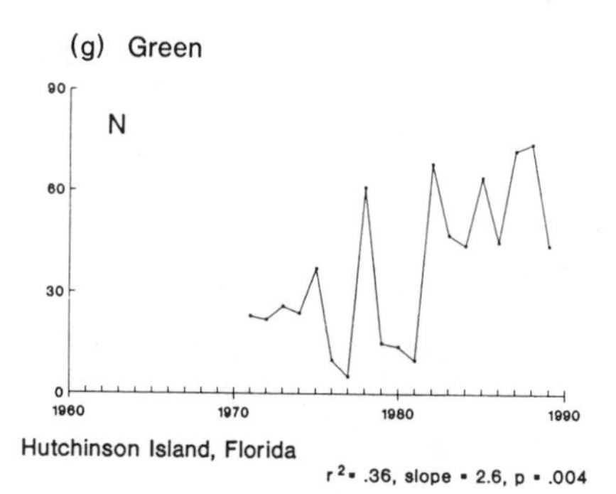

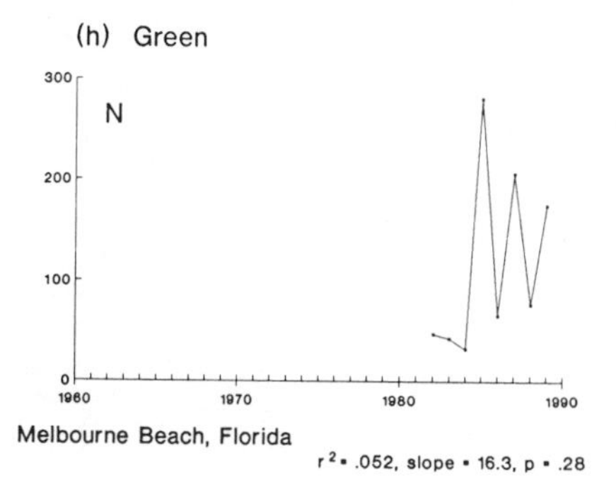

FIGURE 3-1
(Continued)

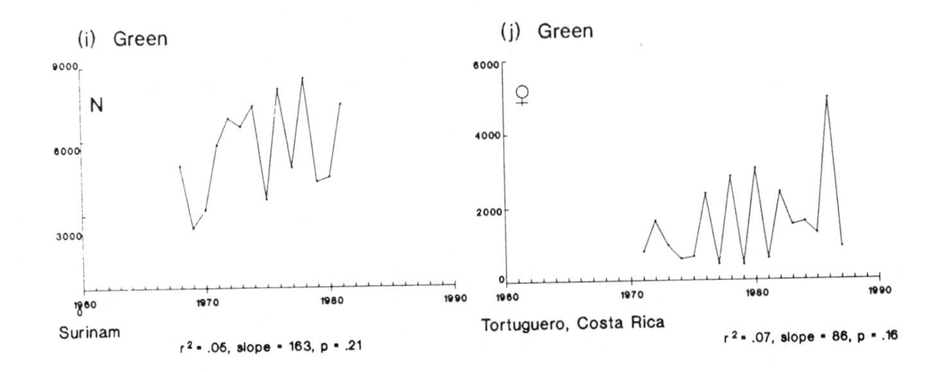

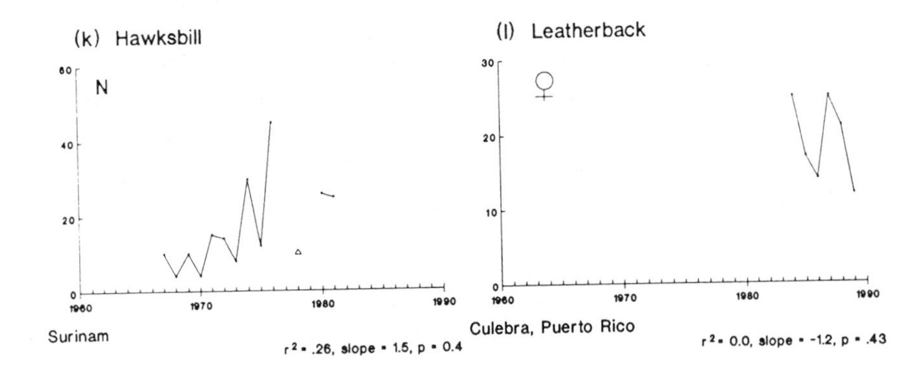

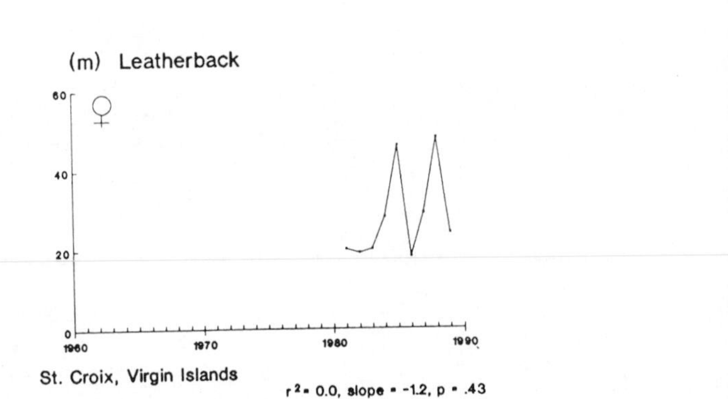

much smaller; about 1,300 females nested on May 31, 1966 (Chávez et al., 1967). Since 1966, the Mexican government, working with the Estación de Biologia Pesquera in Tampico and several other agencies, has maintained a presence on the beach at Rancho Nuevo throughout each nesting season. Personnel have included government turtle biologists, fisheries inspectors, and armed, uniformed Mexican marines. From 1967 to 1970, a few arribadas as large as about 2,000-2,500 turtles were seen (Pritchard and Márquez M., 1973). Archival photographs, probably from 1968, show many hundreds of nesting females on the beach. Over the period 1947-1970, sizes of the largest arribadas on the Rancho Nuevo beach declined dramatically (Figure 3-1*b*).

Since 1978, nests on the Rancho Nuevo beach have been counted by a binational team of Mexican and U.S. scientists working with the U.S. Fish and Wildlife Service. From 1978 to 1988, the number of nests (USFWS Annual Reports, Albuquerque office, 1978-1988) declined significantly (linear regression, $p < 0.05$) by about 14 nests per year (Figure 3-1*a*).

Given that the 1947 arribada estimate was a single count of females on the beach and that a female might be expected to lay 2.3 clutches per season (Pritchard, 1990), the average of about 800 nests per year from 1978 to 1988 would be less than 1% of the estimated nests in 1947 (92,000). This is the most severe population decline documented for any sea turtle species.

LOGGERHEAD

Nesting loggerhead females on Little Cumberland Island exhibit a clear decline in numbers over 26 years (Figure 3-1*d*). The average decline of about 3% per year is not smooth, but the overall downward trend is unmistakable. North of Little Cumberland Island, the number of nests on Cape Island, South Carolina, also shows a decline over 17 years (Figure 3-1*c*). Both populations appeared to undergo a marked decrease in the mid-1970s, but the cause remains unknown. The number of nests per year along the entire South Carolina coast has been estimated from aerial surveys (pers. comm., S. Murphy, S.C. Wildlife and Marine Resources, 1989). Again, a decline is apparent, but year-to-year variability is large. Murphy and Hopkins-Murphy (1989) summed 3-year counts (1980-1982 and 1987-1989) and compared the results; the comparison showed a 26% statewide decrease. The same declining trend was evident for the northern and southern portions of the state and for developed and undeveloped beaches.

About 90% of U.S. loggerhead nesting occurs in Florida from the Canaveral area southward (Hopkins and Richardson, 1984). The important nesting beaches south of Cape Canaveral do not show the declines in nesting characteristic of Georgia and South Carolina. Combined data for 12 years from nine 1.25-km study sites on Hutchinson Island show a possible rising trend in numbers of nests from 1973 to 1989 (Figure 3-1*f*). The most important loggerhead nesting beach in the United States, near Melbourne Beach, Florida (Jackson et al., 1988), has been surveyed for only 8 years (Figure 3-1*e*); no clear trend is apparent.

No important nesting has been observed over the roughly 200 km from New Smyrna Beach to Jacksonville Beach; this gap constitutes some evidence of discrete northern and southern U.S. populations, an idea supported by morphometric differences (Stoneburner et al., 1980) and recently reported genetic differences (pers. comm., B. Bowen, University of Georgia, April 1990). If the separation is genuine, trend data indicate a decline in loggerhead populations from the northern nesting assemblage, but no decline or a possible increase in the southern assemblage. More years of nesting and data and population biology studies are needed to assess trends in the southern assemblage.

GREEN TURTLE

The status and history of green turtle nesting in Florida have been reviewed by Dodd (1982), who found little evidence of past large-scale nesting in the area. Thus, current nesting rates cannot be compared with historical records. The numbers of nests have increased on Hutchinson Island over the period 1971-1989 (Figure 3-1*g*). Considerable nesting also occurs on Melbourne Beach (Figure 3-1*h*), but nests have been counted for only 8 years, a period that is not long enough to confirm a trend.

Green turtles exhibit wide year-to-year fluctuations in numbers of nesting females, and that makes statistical analysis of trends particularly difficult. The year-to-year variation is also apparent in green turtle nesting data from Surinam and Tortuguero, Costa Rica (Figure 3-1*i,j*). The only other substantial regional nesting population, on Aves Island, Venezuela, has not been surveyed long enough for determination of trends, although qualitative observations during visits over many years suggest a heavy decline (Pritchard and Trebbau, 1984).

HAWKSBILL

The hawksbill is an exceedingly difficult species to monitor for long-term trends, for a number of reasons. Small numbers of animals nest on a wide variety of beaches across a broad geographic area. Hawksbill beaches tend to be remote, inaccessible, and sometimes so narrow that the turtle leaves no crawl trace. Hawksbills also exhibit the large year-to-year fluctuations in nesting counts characteristic of green turtles and loggerheads. Thus, few trend data are available. Nests on Buck Island in the Virgin Islands (Hillis and Mackay, 1989b) and Long Island in Antigua (Corliss et al., 1989) have been counted accurately only for the past few years. Mona Island, Puerto Rico, is a concentrated nesting area that has proved to be too remote for consistent assessment (pers. comm., J. Richardson, University of Georgia, 1989). A survey of nests in Surinam (Figure 3-1*k*) has provided a series of 13 annual estimates over 15 years. The trend is positive, but the small number of turtles and the absence of recent data make the trend questionable.

LEATHERBACK

Leatherbacks do not nest with enough frequency on the U.S. mainland (Florida) to permit a trend analysis, although they occur commonly off shore. The nesting beaches nearest the U.S. mainland are those at St. Croix in the Virgin Islands and Culebra, Puerto Rico (Figure 3-1*l,m*). The short records (9 and 6 years) do not indicate trends. Most leatherbacks in U.S. coastal waters are thought to come from Surinam and French Guiana nesting beaches (pers. comm., P. Pritchard, Florida Audubon Society, 1989). Nests on those beaches have been counted since 1967, but the results (as an indicator of population trends) are questionable, because the nesting population has apparently been shifting between the two countries (pers. comm., P. Pritchard, Florida Audubon Society, 1989). Similarly, the small nesting populations in Trinidad and Guyana in the 1960s showed a significant increase by the 1980s, although again a shift from the major beaches in French Guiana cannot be ruled out.

SUMMARY

The committee concluded that population trends are often challenging to interpret, and adequate surveys spanning 10 years or more are usually required to demonstrate with some certainty a change in absolute population numbers. However, much can be deduced about sea turtle trends from the studies of nesting densities to date.

- The Kemp's ridley population has experienced a major decline since 1947, and in the last decade its numbers have continued to decrease.

- Loggerhead nesting populations have declined over the last 20-30 years on northern U.S. nesting beaches (Georgia and South Carolina). On southern Florida Atlantic beaches, however, loggerheads have not shown a decline, and might even be increasing.
- Green turtle nestings on Florida beaches are low but are increasing at Hutchinson Island, Florida.
- Hawksbill nesting is too sparse in U.S. waters for trend analysis. Nesting in Surinam appears to have increased somewhat over the last 15 years, but absolute numbers have been very low throughout.
- Leatherbacks nest in small numbers in the United States, principally in the Virgin Islands and Puerto Rico. Although records are too few to detect trends, the numbers do not appear to be declining. Interpretation of trends on the important Surinam and French Guiana beaches is complicated by population shifts as beaches erode and accrete.

4
Distribution of Sea Turtles in U.S. Waters

To understand the issues concerning the conservation of sea turtles in U.S. waters, we need to view their distribution along the Atlantic and gulf coasts on a broad spatial scale. That immediately makes apparent the wide extent of the complex conservation problem even in U.S. coastal waters. It also helps to identify, for example, which beaches should receive priority for protection of sea turtle nesting and where the distribution of sea turtles overlaps with human activity to cause mortality along the coasts at various water depths in different seasons. This chapter enlarges the general presentation on species distributions in Chapter 2 and provides a broad analysis of the distribution of sea turtles in U.S. waters in recent years. For our analysis, we have taken the most quantitative published information available or have reanalyzed the most extensive data bases available through the cooperation of individuals and government agencies.

SOURCES OF INFORMATION

Nesting Distribution

Information on distribution of nests of loggerheads, green turtles, and leatherbacks in the continental United States has been obtained from aerial surveys and beach patrols. The committee's compilations are based on data from the U.S. Fish and Wildlife Service, North Carolina Wildlife Resources Commission, South Carolina Wildlife and Marine Resources Department, Georgia Department of Natural Resources, and Florida Department of Natural Resources. Additional data were obtained from the U.S. Recovery Plan. Density (nests per kilometer) varies from year to year, as does the intensity of beach surveys. Sufficient data are available, however, to indicate the general density of nesting on beaches from Maine to Texas.

Pelagic Aerial Surveys

Aerial surveys documenting the distribution of sea turtles in the water have been conducted from Maine to the Mexican border. Data presented here are from N.B. Thompson (pers. comm., NMFS, 1989) and Winn (1982). Aerial surveys are valuable for surveying large areas in a short time. However, interpreting data from aerial surveys is difficult for several reasons: small turtles, particularly Kemp's ridleys, generally are not visible, and ocean conditions, such as water clarity and surface glare, can alter visibility and therefore affect the reliability of species identification and counts.

Sea Turtle Strandings

Volunteers in the Sea Turtle Stranding and Salvage Network (STSSN) attempt to document every sea turtle stranding on the U.S. Atlantic and gulf coasts. The date and location of each stranded turtle are recorded, as well as its species, size, and condition. Distribution of strandings provides information on the distribution of turtles. However, quantification of turtle distribution based on that data base is limited by several factors. First, the data base is not independent of the distribution of human-induced mortality factors, such as fishing, dredging, and boating. Second, temporal and spatial coverages are rarely uniform. Most beaches are surveyed by volunteers. Areas under contract for regular surveys since 1986 are fishing zones 17-21 (Texas), fishing zones 4 and 5 (gulf coast of south Florida), and fishing zones 28-32 (Atlantic coast of north Florida, Georgia, and South Carolina) (Figure 4-1). Shorelines formed by marsh or mangrove stands, such as large sections of the Louisiana coast and the northwestern coast of the Florida peninsula, are not surveyed. Third, because of current and wind patterns, dead turtles might float some distance before they strand or might never strand.

FIGURE 4-1 Shrimp-fishing zones along U.S. coasts of Atlantic Ocean and Gulf of Mexico.

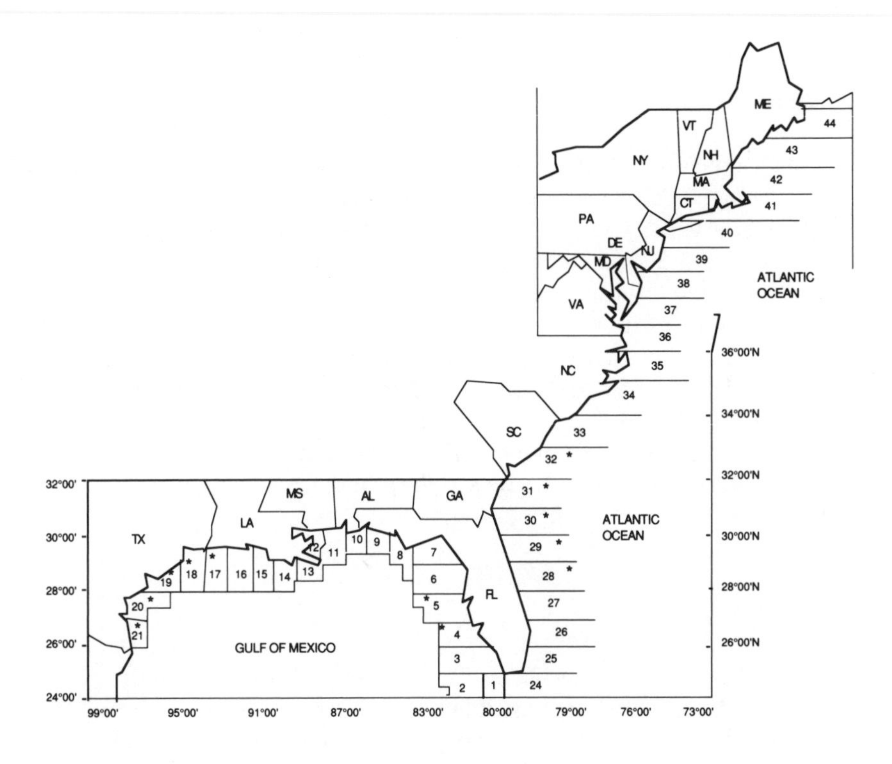

NOTE: Asterisk indicates that zone has contractual arrangement for observing turtle stranding.

DISTRIBUTION

The capture of sea turtles in bottom trawls associated with commercial and experimental or exploratory fishing provides some information on depth and area distribution.

Nesting

The southeastern United States supports one of the two largest rookeries of loggerheads in the world. Some nesting occurs from North Carolina to Louisiana, with outliers as far north as New Jersey and west to Texas (Figure 4-2, *top*); but the 330 km of beach on the Atlantic coast of Florida between St. Augustine and Jupiter supports by far the highest density of loggerhead nesting (Figure 4-2). In recent years, from 50 to more than 200 nests/km of beach are dug annually in this region, compared with only a few to 50 nests/km elsewhere (Figure 4-2, *top*). In addition, the same 330 km of beach is the only location where substantial (but much lower) numbers of green turtles and leatherbacks nest on the U.S. Atlantic and gulf coasts. Kemp's ridleys and hawksbills very rarely use U.S. continental beaches for nesting.

Aerial Surveys

Quantitative data are available for some regions to evaluate the seasonal changes in on/offshore distribution or the depth distribution of large individuals of the most abundant species, the loggerhead, along the Atlantic and gulf coasts of the United States. The most general picture comes from distributional maps compiled from aerial surveys taken in each quarter of the year for much of the Atlantic coast and portions of the gulf coast (Winn, 1982; Thompson, 1984; pers. comm., N.B. Thompson, NMFS, 1989). Other aerial surveys are more spatially restricted but provide useful information for selected sites off Florida, Louisiana, and Texas (Fritts and Reynolds, 1981; Fritts et al., 1983; pers. comm., R. Lohoefener, NMFS, 1989).

North of Cape Hatteras to the Gulf of Maine, large loggerheads were sighted from inshore to the offshore banks and shelf edge and continental slope (Winn, 1982). The distribution shifted from more inshore to more midshelf from spring to summer. From Cape Hatteras, North Carolina to St. Augustine, Florida, sea turtles, mostly large loggerheads with a few adult leatherbacks, generally appeared more abundant on the inshore

FIGURE 4-2 Distribution of loggerhead nesting (per km) and seasonal aerial surveys of loggerheads (per 10,000 km²) in shrimp-fishing zones. Data from Appendix D.

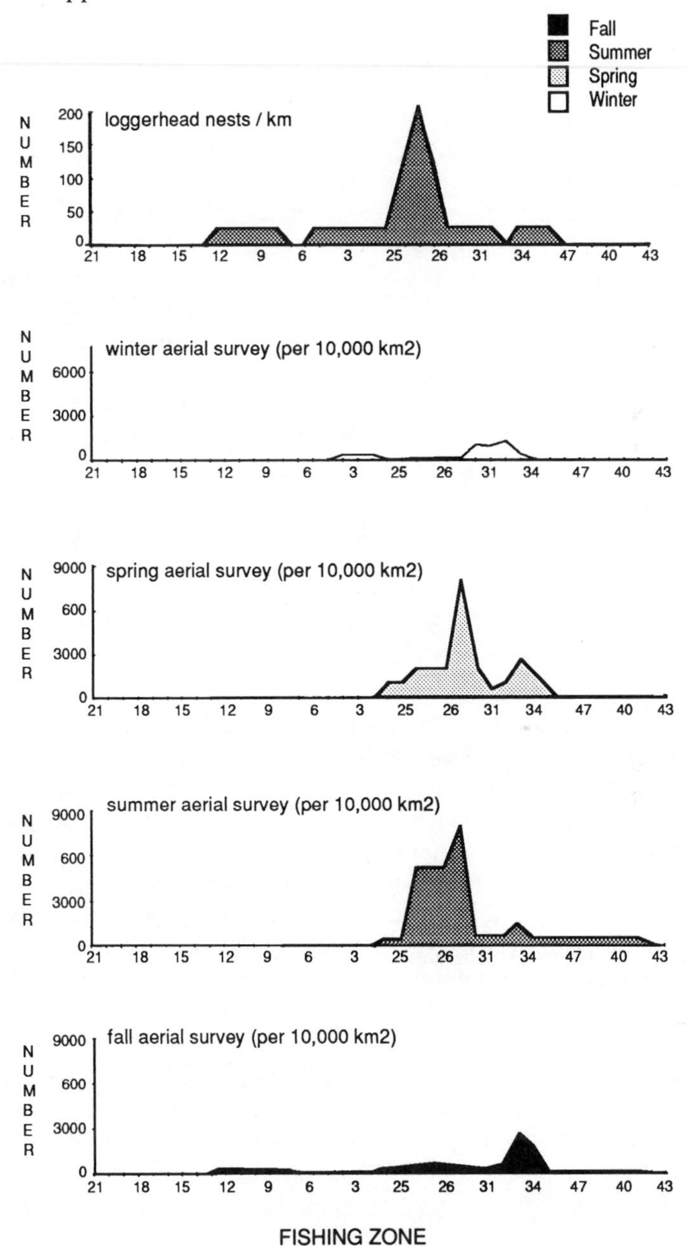

halves of aerial transects than on offshore halves in spring and summer, but appeared less abundant on the inshore than offshore halves in fall and winter (Thompson, 1984). There are too many points on Thompson's maps to see any obvious difference in the Cape Canaveral region. South of Canaveral, large loggerheads appear more abundant in the inshore than offshore halves of the transects in all seasons of the year. In the Gulf of Mexico from Key West to the Mississippi River (pers. comm., N. Thompson, NMFS, 1989), sightings of large loggerheads seem more frequent in the inshore portions of aerial surveys than in the offshore portions in summer and autumn, and offshore in winter. Maps of the sightings of large loggerheads used in Lohoefener et al. (1988) in spring and autumn for all gulf locations show no obvious seasonality with respect to distance from shore, nor did Lohoefener (pers. comm., NMFS, 1989) observe any seasonal changes in depth distribution off Louisiana from the data used by Lohoefener et al. (1989).

Densities of large loggerheads (with a few adult leatherbacks) from aerial sightings can also be analyzed with respect to water depth over which the turtles were sighted within survey areas of 25,642 km^2 at two locations on the gulf coast of southern Florida in August (Fritts and Reynolds, 1981) and seasonally both for the Atlantic coast of Florida off the primary nesting beaches of loggerheads near Cape Canaveral and the gulf coast of southern Florida (Fritts et al., 1983). Other sites off Louisiana and Texas had too few turtle sightings to analyze for seasonality of on/offshore or depth distributions.

The primary conclusion of these two aerial surveys off Florida is that both in the Atlantic waters (Canaveral area) and the gulf waters of southern Florida, the aerial sighting densities of large loggerheads are higher throughout the year over water depths of 0-50 m than over depths from 50-1,000 m; few large loggerheads or leatherbacks were observed over waters from 50-1,000 m in any season. Averaged over all seasons, the sighting densities over waters 25 to 50 m deep were 78-82% of those over 0 to 25 m depths, but sighting densities over waters 50 to 100 m deep were 9-14% of those over 0 to 25 m depths. Because the depth contours drop off much more sharply at Canaveral than at the gulf site off south Florida, it also appears that the distributions of large loggerheads were related to water depth rather than to distance from shore. For both locations, the sighting density declines rapidly near the 50 m depth contour rather than at a fixed distance from shore. An alternative explanation might be that turtles spend more time below the surface in deeper water and that fewer are then sighted. However, the catch in trawls, presented below, also supports the conclusion of fewer large loggerheads and leatherbacks being found in deeper waters.

At both the Canaveral and the southern Florida gulf sites, large loggerheads remained abundant throughout the year at depths from 0 to 50 m. In waters less than 50 m deep, minimum sighting densities of large loggerheads observed in October and December averaged about 50% of those for February, April, June, and August.

Aerial surveys of coastal waters also demonstrate the high concentration of adult loggerheads off the primary nesting beaches along the Atlantic coast of Florida during spring and summer (Figure 4-2); sightings range up to about 7,900 per 10,000 km^2. Moderately high sighting densities, about 2,500 per 10,000 km^2, also were reported in the fall off North and South Carolina. Densities of sighted large loggerheads were low (about 30-100 per 10,000 km^2), along portions of the west coast of Florida, and decreased sharply off Louisiana and Texas (to 1-30 per 10,000 km^2). North of Cape Hatteras, loggerheads were absent in winter, low in summer (about 500 per 10,000 km^2, and very low (1-4 per 10,000 km^2) even in summer as far north as the Gulf of Maine.

Leatherbacks sighted in aerial surveys were uncommon throughout the entire Gulf of Mexico, averaging about 50 per 10,000 km^2 (Lohoefener et al., 1988) and were about one-hundredth as abundant as large loggerheads among identified sightings off the Atlantic coast south of Cape Hatteras (Thompson, 1984). In the Gulf of Maine, leatherbacks numbered only 7-8 per 10,000 km^2 during summer and fall; they were absent or very sparse in winter and spring.

Kemp's ridleys are not usually visible and identifiable from aerial surveys, so this survey method provides no information on their distribution.

Seasonality of sighting densities varies with the geographic location along the coast. Off the primary nesting beaches of Florida's Atlantic coast, sighting densities were about 15 times higher during spring and summer than during autumn and winter (Figure 4-2); the lowest sighting densities occurred in the winter, when they were about 2.5% of highest summer densities. That pattern reflects the aggregation of the mature loggerheads for breeding and access to the nesting beaches. Farther north, off North Carolina, sighting densities were not maximal during the summer nesting season, but rather were 2-4 times higher during spring and autumn than during winter or summer. Seasonal coverages of aerial surveys are insufficient to permit speculation about other regions.

Strandings

According to 1987 and 1988 data from the STSSN, the most common turtle carcasses found on the outer beaches from Maine to Texas were those of loggerheads (1,522 and 1,150 in these years), followed by Kemp's ridleys (141 and 176), green turtles (105 and 150), leatherbacks (119 and 63), and hawksbills (22 and 20) (Appendix E). Those numbers understate the number of dead turtles in the area, in that many dead turtles do not drift ashore or are not found. The highest stranding rates of loggerheads occurred along 500 km of Atlantic beaches of Georgia and northern Florida (Figure 4-3). Other areas with many strandings of loggerheads were the beaches of Mississippi, Alabama, and Texas. Carcasses of Kemp's ridleys were found most frequently on beaches of Texas, the Atlantic coast of northern Florida, and North Carolina (Figure 4-3). Green turtles were stranded most frequently along the Atlantic coast of Florida; leatherbacks along the coasts of Delaware, New Jersey, and New York; and hawksbills along the coasts of Texas and Florida (Figure 4-3).

Seasonality of strandings differs with species of turtle and geographic region. Loggerheads strand most frequently in May-December on Atlantic beaches, in April and May on the Texas coast, and in May and June in Mississippi and Alabama. In some locations and seasons, few turtle carcasses are reported. In some areas, that is accounted for by the absence of beach surveys or by ocean current patterns; in others, it might be related to an overall lack of turtles in the region. For example, in northern Florida and Georgia, only 1.5% and 3.6% of the annual totals of loggerhead strandings in 1987 and 1988, respectively, occurred in winter (January-March) along the 500 km characterized by maximal strandings during May through September. That is consistent with the aerial survey data on turtles off this coastal region, where winter sighting densities were 2.5% of maximal summer sighting densities. In addition, few sightings in the region were on the inshore portions of the aerial surveys in winter, but many during the spring and summer (Thompson, 1984).

Seasonality of strandings of Kemp's ridleys also appeared to differ with region (Figure 4-3). On Texas beaches, stranding occurred in February-December, with maximums in April and May and again in August and September, but few strandings occurred on the Atlantic coasts of Florida to Maine in January-May.

ONSHORE, OFFSHORE, AND DEPTH DISTRIBUTION

Turtles caught in bottom trawls also provide information on depth distribution that is consistent with the marked decrease of large loggerheads and leatherbacks at increasing depths observed in the aerial surveys. Twenty-nine loggerheads were captured off Georgia and Florida (to Key West) in 1306 hours of trawling (Bullis and Drummond, 1978). The highest catches, about 0.0015-0.0045 turtles/hour of trawling, were taken in

FIGURE 4-3 Sea turtle stranding, by species, 1987-1988. Fishing zones are shown on horizontal axes (see Figure 4-1). Source: STSSN (see Appendix E).

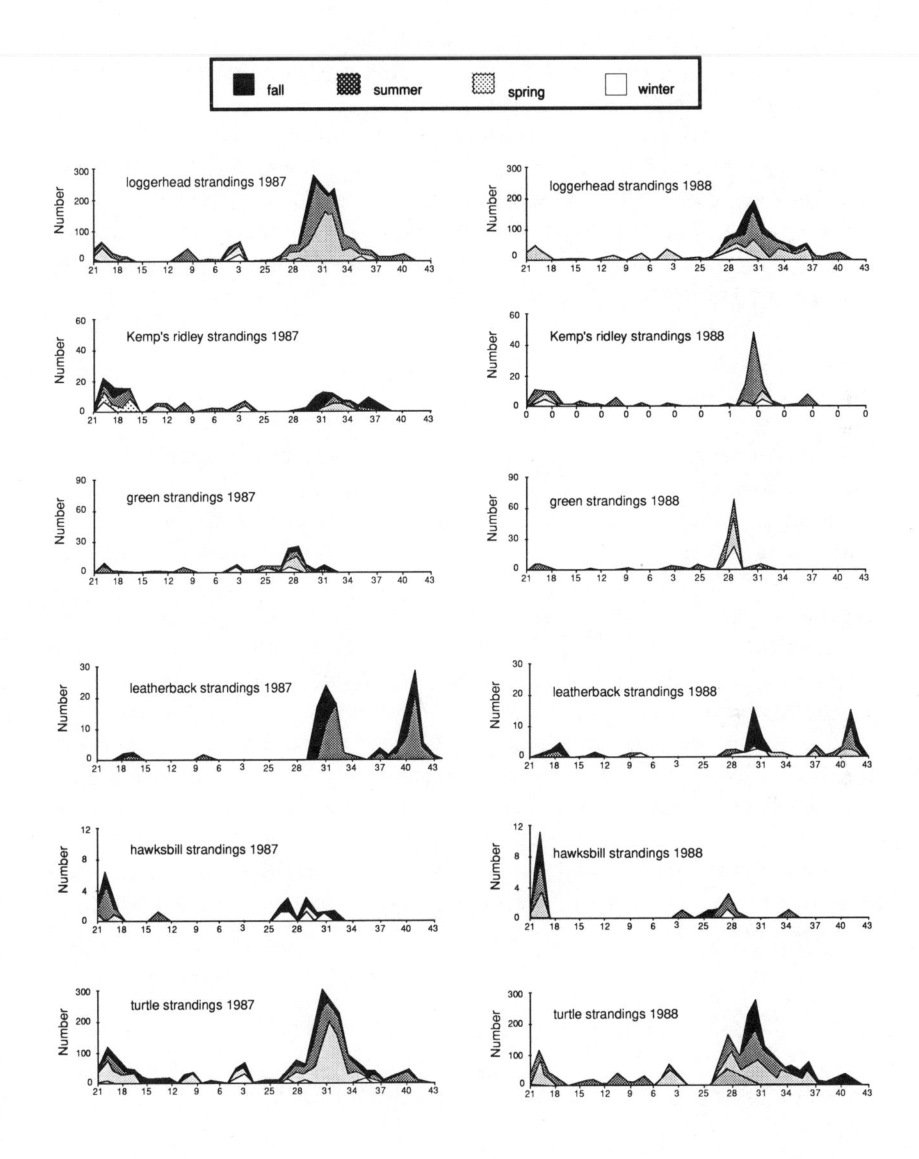

0-40 m of water compared with catch rates of 0-0.0025 turtles/hour in 40-100 m. Thirteen sea turtles, mostly loggerheads and Kemp's ridleys, were caught in the Gulf of Mexico off Louisiana or the west coast of southern Florida in the NMFS observer program in 1988-1989. Catch rates were 0.006 turtles/hour (in 2,007 hours) at 0-27 m and 0.0008 turtles/hour (1,285 hours) at more than 27 m. Henwood and Stuntz (1987) also showed that the catch of sea turtles per net per hour was lower at depths of 27-99 m than 2-27 m. They had 976 trawling hours in the deeper water and 5,177 trawling hours in the shallowest water. They did not present data on catch rates of turtles by depth, but the catch per effort at depths greater than 27 m was less than the catch per effort for all but two of the other depth intervals. As with the aerial surveys, turtle abundance in deeper water appeared to be about one-tenth that in shallower water.

SUMMARY

Data on distribution of sea turtles come from observations of nesting turtles, aerial surveys, the STSSN, and incidental captures in fishing gear.

Nesting is most common on the Atlantic coast of Florida, and the loggerhead is the greatly predominant species. Loggerheads aggregate off the nesting beaches in spring and summer, and move up and down the coasts and a little more offshore in fall and winter. Some leatherback and green turtle nesting occurs in eastern Florida. According to stranding data, sea turtles in order of decreasing abundance in U.S. coastal waters are loggerheads, Kemp's ridleys, green turtles, leatherbacks, and hawksbills. Some strandings occur from the Gulf of Maine to all the states along the Gulf of Mexico. Adult turtles are apparently less abundant in deeper waters of the Gulf of Mexico than in waters less than 27-50 m deep, and they are usually uncommon near the shore off northern Florida in fall and winter. In the eastern gulf, turtles are less abundant inshore in winter than in summer, but even in winter they are common in inshore waters.

5
Natural Mortality and Critical Life Stages

This chapter summarizes current information on the causes and magnitude of natural mortality of sea turtles, and discusses how sea turtles at different life stages contribute to the population or to the reproductive value. Recent analyses of loggerhead populations and reproduction (Crouse et al., 1987) are especially useful for making decisions about conservation of sea turtles, because they help to identify life stages in which reduced mortality can have the greatest influence on the maintenance or recovery of endangered or threatened sea turtle populations.

From models developed by Frazer (1983a), female loggerheads probably first nest when about 22 years old, and survivors continue nesting every few years until they are about 54. Most mature female loggerheads nest every second or third year and deposit several clutches of eggs during a nesting season. Thus, an individual is estimated to lay on the average 80 eggs each year for 30 years. The eggs and hatchlings have high mortality rates, but as the survivors grow, natural mortality declines markedly. About 80% of the nesting females studied for many years at Little Cumberland Island survive from one year to the next. (Chapter 2 presented variations on the pattern of life history of the several species of sea turtles.) These general patterns of mortality and reproduction form a basis for the insight needed to devise a rational program for sea turtle conservation (Crouse et al., 1987).

Sea turtles are killed by various animals and environmental phenomena. Nests and eggs are destroyed by predators, erosion, and inundation by rain or tides. After hatching, turtles of all ages, both at sea and on land, are consumed by predators. They are also subject to debilitating parasites and diseases and are killed by various abiotic factors, including hurricanes and thermal stress. However, quantitative accounts of sea turtle mortality in the wild are few.

Some of the apparently natural factors that are lethal to sea turtles are associated with human activities. For example, sea turtles are subject to predation by wild, formerly domestic animals introduced by humans (hogs and dogs) or wild, nondomesticated animals introduced by humans (mongoose) or enhanced by human activities (raccoons). Beach erosion is a natural source of mortality that has also been altered by human activities.

BIOTIC SOURCES OF MORTALITY

Predation

Many species, from ants to jaguars, prey on sea turtles. Excellent reviews (which include lists of predators) by Hirth (1971), Stancyk (1982), Witzell (1983), and Dodd (1988) categorize predators by the life stage of sea turtles on which they prey, and the following presentation below follows that pattern.

Eggs and Hatchlings on the Beach

Predators of the Kemp's ridley at Rancho Nuevo, Mexico, include coyotes, raccoons, coatis, skunks, ghost crabs, and ants (Márquez M. et al., 1989). Some predators, such as the black vulture, feed on eggs from nests already opened by other predators or erosion. Hatchling Kemp's ridleys are caught and eaten on the beach by ghost crabs, vultures, grackles, caracaras, hawks, coyotes, raccoons, skunks, coatis, and badgers (Márquez M., in prep.).

The major loggerhead egg predator in the southeastern United States is the raccoon (Dodd, 1988). Before protective efforts were initiated, raccoons destroyed nearly all the nests at Canaveral National Seashore, Florida (Ehrhart, 1979), and at Cape Sable, Florida, raccoons destroyed 85% of the nests in 1972 and 75% in 1973 (Davis and Whiting, 1977). The high rate of predation might have resulted from the unusually large raccoon populations, which were augmented by such human activities as habitat alteration, food supplements (garbage), and removal of natural predators of the raccoon (Carr, 1973; pers. comm., L. Ehrhart, University of Central Florida, 1989). Not all nesting beaches in Florida suffer such high losses from raccoons; for example, only seven of 97 nests on Melbourne Beach, Florida, were destroyed by raccoons in 1985 (Witherington, 1986). Other nest predators are ghost crabs, hogs, foxes, fish crows, and ants (Dodd, 1988). From 1980 to 1982, nonhuman predators destroyed up to 80% of the loggerhead clutches laid on two barrier islands in South Carolina (Hopkins and Murphy, 1983).

Management practices have eliminated nearly all the beach predation of Kemp's ridleys at Rancho Nuevo, and reduced predation significantly on most of the important loggerhead nesting beaches.

Hatchlings as They Leave the Beach

Once in the ocean, Kemp's ridley hatchlings are eaten by a large variety of predatory birds and fish (Márquez M., in prep.). Loggerhead hatchlings at this time in their lives also fall prey to a similar array of predators, including gulls, terns, sharks, and other predatory fish (Dodd, 1988). Many Atlantic sharpnose sharks captured in a commercial fishery off Florida during the turtle hatching season in 1988 had loggerhead hatchlings in their stomachs (pers. comm., A. Bolten, University of Florida, 1989).

Larger Juveniles and Adults in the Water

Sharks and other large predatory fish are important predators of Kemp's ridleys in all oceanic life stages (Márquez M. et al., 1989). Tiger sharks might be selective predators of large cheloniid sea turtles; analyses of stomach contents of 404 tiger sharks showed that 21% of the sharks with food in their stomachs had eaten large turtles (Witzell, 1987). Balazs (1980) has summarized data on predation of juvenile and adult green turtles in Hawaii by tiger sharks; turtles were found in 7-75% of tiger sharks sampled in Hawaiian waters inhabited by sea turtles.

Nesting Females on the Beach

There is no evidence of nonhuman predation of adult loggerhead females on U.S. nesting beaches, but it might have occurred in the past.

Reported predators of leatherbacks, green turtles, and hawksbills are similar to those of loggerheads and Kemp's ridleys at each life history stage (Hirth, 1971; Pritchard, 1971; Fowler, 1979; Balazs, 1980; Stancyk,

1982; Bjorndal et al., 1985; Witzell, 1987). The actual predator species change with geographic region, but are from the same feeding guilds.

Diseases and Parasites

Most reported diseases in sea turtles have been described in captive animals (Kinne, 1985). Diseases induced by stress or improper diet in captivity and not known to occur in wild sea turtles (Glazebrook, 1980; Kinne, 1985; Lauckner, 1985) will not be discussed here. An excellent review of the diseases and parasites of sea turtles can be found in Lauckner (1985), and specific parasites of sea turtles are identified in the reviews by Hirth (1971), Witzell (1983), and Dodd (1988).

Cutaneous fibropapillomatosis, a disease of green turtles, has been recorded infrequently in Florida waters for many years (Smith and Coates, 1938). However, large numbers of green turtles have recently contracted the disease in the Indian River lagoon system in east-central Florida (Witherington and Ehrhart, 1989b) and the Hawaiian Islands (Balazs, 1986). In the Indian River, 40-52% of the green turtles captured in 1983-1988 had fibropapillomas. In Hawaii, 10% of the nesting females at French Frigate Shoals had fibropapillomas, as did 35% of 51 stranded green turtles in 1985 (Balazs, 1986). Recaptured turtles have demonstrated further proliferation of the fibropapillomas, although in other cases regression occurs (Witherington and Ehrhart, 1989a). Tumors can cause mortality indirectly. Turtles whose vision is blocked by tumors are unable to feed normally, and turtles with fibropapillomas are more prone to entanglement in monofilament line and other debris (Balazs, 1986; Witherington and Ehrhart, 1989a). Research on the cause of the disease is in progress (Jacobson et al., 1989).

Spirorchidiasis has been reported in loggerheads (Wolke et al., 1982). Severe infestations of spirorchids (blood flukes) result in emaciation, anemia, and enteritis, or conversely, emaciation and anemia could make a turtle more susceptible to spirorchid infestation. Three genera of blood flukes were identified in 14 of 43 loggerheads stranded or floating dead from Florida to Massachusetts (Wolke et al., 1982). Spirorchidiasis can result in death or make turtles more susceptible to succumb to other stresses (Wolke et al., 1982).

A macrochelid mite (*Macrocheles* sp.) has been found on Kemp's ridley hatchlings emerging from relocated nests (Mast and Carr, 1985). Mites of the same genus, considered to be nonparasitic, were found on loggerhead hatchlings in South Carolina (Baldwin and Lofton, 1959).

Bacterial and fungal infections of eggs can be a major source of mortality. Bacteria and fungi are implicated as a major cause of death of olive ridley eggs at Nancite, Costa Rica, where hatching success averages only 5% (Cornelius, 1986; Mo, 1988). Microbial pathogens are believed to cause mortality of loggerhead embryos (Wyneken et al., 1988).

Other Nesting Turtles

Eggs and emerging hatchlings are sometimes killed when their nest is dug into by a nesting female of either the same or a different species. Bustard and Tognetti (1969) described this activity as a density-dependent mortality factor. Although a thorough study of the relationship between nesting density and this mortality factor has not been carried out, clearly the greater the number of nesting females in a given area, the greater the likelihood of a female disturbing an earlier nest. In most areas, this is a minor source of mortality because most nesting populations have densities that are relatively low. However, during the mass nestings (arribadas) of olive ridleys, large numbers of nests can be destroyed. Cornelius (1986) estimated that 7% of the nests of the olive ridley colony at Nancite, Costa Rica, were destroyed by other females' digging in the same arribada, and another 10% were destroyed by females' digging in subsequent arribadas. In contrast, at Tortuguero, Costa Rica, of 587 green turtle nests monitored, none was destroyed by nesting activities of other turtles (Fowler, 1979). At Mon Repos, Australia, an average of 0.43% of the total seasonal egg production in five consecutive seasons was destroyed by nesting loggerheads (Limpus, 1985).

Vegetation

Although usually a minor cause of death, plant roots can invade turtle nests and cause mortalities. Invasion by roots of beach morning glory (*Ipomoea pes-caprae*) and sea oats (*Uniola paniculata*) killed 275 embryos in three of 97 loggerhead nests on Melbourne Beach, Florida (Witherington, 1986), and at Cape Romaine, South Carolina, 5% of the eggs laid among sea oats were destroyed by the roots (Baldwin and Lofton, 1959). Destruction of marine turtle nests by sea oat roots also has been reported by Raymond (1984).

Plants can also entrap sea turtles. Hatchlings get entangled on their way to the sea (Limpus, 1985) and adult females sometimes become fatally trapped in vegetation or by logs washed onto the nesting beach (Pritchard, 1971; Cornelius, 1986).

ABIOTIC SOURCES OF MORTALITY

Erosion, Accretion, and Tidal Inundation

In almost every nesting colony, some nests are lost to erosion, accretion, and tidal inundation. The extent of mortality varies widely among beaches, years, and species. Nests deposited on shifting beaches are more susceptible to destruction from erosion or accretion. In each species, some turtles deposit nests below the high-tide line. Leatherbacks often nest in areas vulnerable to erosion or inundation: 40-60% of the nests in Surinam were in such areas, compared with 12% of green turtles on the same beach (Whitmore and Dutton, 1985), the Guianas, and St. Croix (Eckert, 1987), but less than 3% in Malaysia (Mrosovsky, 1983). Erosion and inundation destroyed 3-25% of the loggerhead nests deposited each year on two barrier islands in South Carolina in 1980-1982 (Hopkins and Murphy, 1983), and on Melbourne Beach, Florida, 17 of 97 loggerhead nests in 1985 were lost to erosion, accretion, and surf action (Witherington, 1986).

Heavy Rains

Heavy rain can destroy large numbers of nests. Ragotzkie (1959) reported that all embryos in 15 of the 17 loggerhead nests deposited on Sapelo Island, Georgia, in 1955 and 1957 were drowned by heavy rain. Kraemer and Bell (1978) also reported heavy loggerhead egg and hatchling mortality in Georgia resulting from heavy rains. At Tortuguero, Costa Rica, heavy rains and high groundwater drowned all embryos in many green turtle nests in 1986 and 1988 (Horikoshi, 1989).

Thermal Stress

Hypothermia in sea turtles causes a comatose condition and can result in death. Perhaps the best-documented events are those that occurred in recent years in Long Island Sound, New York (Meylan and Sadove, 1986), and in the Indian River lagoon system, in Florida (Wilcox, 1986; Witherington and Ehrhart, 1989b). Both areas can act as natural "traps," because of their geographic configurations (Witherington and Ehrhart, 1989b). Of 52 turtles (41 Kemp's ridleys, nine loggerheads, and two green turtles) stranded in Long Island Sound in the winter of 1985-1986, 18 were alive when discovered and 11 (nine ridleys, one loggerhead, and one green turtle) survived after gradual warming at rehabilitation centers (Meylan and Sadove, 1986).

Morning surface water temperatures below 8°C in 1977, 1978, 1981, 1985, and 1986 caused hypothermic stunning of sea turtles in the Indian River lagoon system, in Florida (Witherington and Ehrhart, 1989b). Those events involved 342 green turtles (25-75 cm SCL), 123 loggerheads (44-91 cm), and two Kemp's ridleys (55-63 cm). Among the stranded turtles, a greater proportion of green turtles than of loggerheads died, and smaller turtles were more susceptible to hypothermia. Most of the turtles were released alive, and many were recaptured months or years later. We have no way of estimating the mortality that would have occurred without human intervention (Witherington and Ehrhart, 1989b).

QUANTITATIVE STUDIES OF NATURAL MORTALITY

The only life stage for which natural mortality of sea turtles has been quantified is the egg and hatchling stage, including the brief period when hatchlings emerge from the nest and make their way down the beach to the water. Percentage of emergence of hatchlings is measured and reported in the literature in two ways. In the first, egg clutches are marked as they are laid and followed through the season; that results in an emergence percentage for eggs in all clutches laid. In the second, the emergence success of hatchlings from clutches that successfully produce hatchlings is determined. The former value is the best measure of survivorship. Results in Table 5-1 indicate the range of survivorship values for the egg stage. Of necessity, some studies include sources of mortality related to human activities—for example, predation by humans, formerly domestic animals, and wild animals introduced by humans.

The rate of mortality resulting from predation is assumed to be much higher for eggs and very small turtles than for larger turtles, because the lists of predators on eggs and hatchlings are much longer than those of predators on larger juveniles and adults. However, there are no quantitative studies of predation away from the nesting beach, so the assumption, although a reasonable one, has not been tested.

The value of an individual of a particular age or life stage can be stated according to its expected production of offspring, hence the term "reproductive value." Reproductive value is the relative contribution of an individual of a given age to the growth rate of the population (see Mertz, 1970, for a description of reproductive value). The more offspring an individual is expected to produce, the higher its reproductive value. The life stages that we consider below are eggs and hatchlings, small juveniles,

TABLE 5-1 Emergence success of sea turtle egg clutches presented as mean (range). Emergence success is the percentage of eggs that produce hatchlings that reach the surface of the sand above the nest chamber. Data are presented only for natural nests (nests not moved or protected) from studies that included those clutches that produced no hatchlings.

Species and Location	Clutches (Number)	Emergence Success (%)	Reference
Loggerhead			
Tongaland	72	78 (0-99)	Hughes, 1974
Brevard Co., Florida	97	56 (0-99)	Witherington, 1986
Cape Canaveral (1982)	310	1 (0-90)	McMurtray, 1982
Cape Canaveral (1983)	76	3 (0-?)	McMurtray, 1986
Green turtles			
Bigisanti, Surinam	57	84	Schulz, 1975
Hawaii	40	71 (0-93)	Balazs, 1980
Tortuguero, Costa Rica	318	35 (0-?)	Horikoshi, 1989
Florida	25	57 (0-94)	Witherington, 1986
Hawksbill			
U.S. Virgin Islands	61	60 (0-100)	Small, 1982
U.S. Virgin Islands	88	81	Hillis and Mackay, 1989a
Tortuguero, Costa Rica	5	36 (0-94)	Bjorndal et al., 1985
Antigua, West Indies (1987)	99	79 (0-100)	Corliss et al., 1989
Antigua, West Indies (1988)	156	85 (0-100)	Corliss et al., 1989
Leatherback			
Bigisanti, Surinam	52	50	Schulz, 1975
Culebra, Puerto Rico	429	71 (0-100)	Tucker, 1989b
St. Croix (1983)	98	25 (0-95)	Eckert and Eckert, 1983
St. Croix (1984)	123	26 (0-97)	Eckert et al., 1984

large juveniles, subadults, and nesting adults (breeders). The life stage with the highest reproductive value is the one for which greater protection can contribute the most to the maintenance or recovery of a population.

Reproductive value can be estimated with population models. Those models have a long history in population ecology, perhaps beginning with Lotka (1922). The models traditionally combine information on age-specific fecundity and age-specific survivorship to yield population projections where survivorship is the percentage of individuals that survived the year and fecundity is the average number of eggs produced per female. Other important factors in the calculations are the number of years required for an animal to reach its reproductive age and the ratio of females to males in the population.

The concept of a mathematical value for reproductive value arose from Cole's use of demographic models (Fisher, 1958). A reproductive value of 1 is assigned to a newly laid egg, and all other ages receive valuations relative to that. The idea of reproductive value is fundamental to conservation biology, because it helps to identify the age classes of most significance for determining future population size.

Population modeling is also useful in assessing whether a particular population is growing or declining and at what rate. Its greatest usefulness, however, might be in sensitivity analysis (Cole, 1954), the estimation of the magnitude of change in the growth rate of the population for each of several changes in such factors as fecundity and survivorship. A sensitivity analysis can evaluate, for example, whether a 10% increase in survivorship could have the same effect on population growth as a 50% increase in fecundity. If it did, then the growth of the population would be 5 times more sensitive to survivorship changes than to fecundity changes. Sensitivity analysis is also useful for predicting which of several life stages would be most responsive to a particular management tool.

The loggerhead is the sea turtle whose demographics are best known, because loggerheads nest in sufficient numbers along the southeastern U.S. coast to be accessible to scientists, and because one nesting population on Little Cumberland Island, Georgia, has been subject to intensive tagging since 1964 (Richardson and Hillestad, 1978; Richardson and Richardson, 1982). Frazer has conducted an exhaustive analysis of the Cumberland loggerhead population (Frazer, 1983a,b; 1984; 1986; 1987; Frazer and Ehrhart, 1985; Frazer and Richardson, 1985a,b; 1986) and has provided the algebraic notation for the standard age-based population model of Lotka (1922).

Survivorship and fecundity in loggerheads are best estimated by life history stages (eggs, hatchlings, small pelagic juveniles, large coastal juveniles, subadults, and adults), rather than years of age, so Crouse et al. (1987) used Frazer's demographic data from Cumberland Island loggerheads to apply a stage-based demographic technique for analyzing population dynamics. The approach, as developed by Werner and Caswell (1977), is analogous to the traditional age-based life-table analysis, but does not require age-specific information.

Population factors (Table 5-2, columns 1-4) used by Crouse et al. (1987) in the analyses were calculated by Frazer (1983a). Predictions for reproductive value (column 5) and sensitivity (column 6) were derived from the model of Crouse et al. (1987). Five life stages are represented. Annual survivorship is lowest in eggs and hatchlings—67% per year—and in large juveniles—68%. Large juveniles are the dominant size group (55-75 cm) of the turtles stranded on the beaches of North Carolina (Crouse et

TABLE 5-2 Annual survivorship and reproductive value of loggerheads in five life history stages.

Life History Stage	Size (cm)	Approximate Ages (years)	Annual Survivorship (% per year)	Fecundity (eggs/year)	Reproductive Value (relative to egg/ hatchling)	Greatest Benefit from Protection (rank of 1 = highest benefit)
Eggs, hatchlings	<10	<1	67	0	1	5
Small juveniles	10-57	1-7	79	0	1.4	3
Large juveniles	58-79	8-15	68	0	6	1*
Subadults	80-86	16-21	74	0	116	2
Breeders	>87	22-54	81	80	584	4

*This life stage offers the greatest management potential for increasing the future growth of the population. Younger animals have a lower reproductive value, because most will not reach maturity. Older animals have a higher reproductive value, but very few are left in the population to reproduce. Protection of large juveniles has the greatest effect on increasing the future growth of the population.

Source: Modified from Crouse et al., 1987.

al., 1987) and other beaches from Florida to North Carolina (Schroeder, 1987; Schroeder and Warner, 1988; Schroeder and Maley, 1989). Survivorship is estimated to be highest for nesting females—81% per year. Because they do not breed until they are 12-30 years old, and 22-33% die each year, few loggerheads reach reproductive age. The reproductive value of individual surviving turtles is greatest for breeders, which, once they reach maturity, can continue to breed for many years. Each individual breeder's reproductive value is estimated to be about 584 times greater than that of an egg or hatchling. Few turtles, however, survive to adulthood and reproduce. As Crouse et al. (1987) noted, "By increasing the survival of large juveniles (who have already survived some of the worst years) a much larger number of turtles are likely to reach maturity, thereby greatly magnifying the input of the increased reproductive value of the adult stages."

The analyses of Crouse et al. (1987) suggested that the greatest increase in growth rate of the Little Cumberland Island population could be achieved by increasing the survivorship of the large juveniles and subadults. Increasing fecundity or survivorship of eggs had less influence on population growth than increasing survivorship of older turtles. This conclusion was not especially sensitive to uncertainties in the parameter estimates. Because beach strandings of dead sea turtles are dominated by large juveniles (Crouse et al., 1987), reducing strandings would affect the very life stage whose increased survivorship could increase loggerhead population growth the most. No conservation effort can be successful without adequately protecting all stages in the life cycle, but the analyses of Crouse et al. (1987) strongly suggest that efforts to reduce mortality of larger juvenile and adult loggerheads will be more effective at promoting loggerhead population growth than efforts to increase the numbers of hatchlings leaving the beaches. The analyses also predict that efforts to protect eggs on nesting beaches and efforts at "headstarting" loggerheads would by themselves be insufficient to reverse the observed decline in the population of loggerheads nesting on Little Cumberland Island (Figure 3-1*d*).

Although the results of such population models clearly depend upon necessary assumptions regarding poorly known demographic characteristics, the general conclusions of the Crouse et al. (1987) model of the loggerhead are robust. Of the poorly known demographic characteristics, age at sexual maturity is the one to which the model is most sensitive. But large changes in maturation rate and in other imprecisely known demographic characteristics did not alter the general conclusion that increasing the survivorship of juveniles and young adults would promote population growth far more than increasing survivorship of eggs and hatchlings. However, the imprecision of our knowledge of necessary demographic characteristics for loggerheads prevents us from specifying how many hatchlings would have to be spared to equal the effect of sparing the life of a single large juvenile, although we know that the number is large.

Crouse et al. (1987) modeled only the loggerhead, but there are reasons to believe that aggregate reproductive value in Kemp's ridley and other sea turtles is also greater for larger juveniles and young adults than for earlier and later stages. The key demographic characteristics that lead to this pattern in how reproductive value varies with life stage are the relatively long time to sexual maturity and the extremely high mortality rate from birth to age of sexual maturity. Those characteristics ensure that reproductive value of individual hatchlings will be relatively low. To the degree that all sea turtles share those two traits with the loggerhead, the conclusion that reproductive value of hatchlings is relatively low will apply generally. The implication for conservation efforts, too, is general: Increasing survivorship of older juvenile and young adult sea turtles is the most effective means of increasing population sizes. Because mature sea turtles age without ceasing to reproduce, reproductive value will remain high until late in adult life, thus suggesting that continued protection of adult sea turtles will be an important conservation measure. However, if there are few or no hatchlings, there will inevitably be few or no adults ultimately. Therefore, relative reproductive values will be useful in management decisions only if there is a certainty that large numbers of hatchlings are being produced.

SUMMARY

Sea turtles lay great quantities of eggs throughout their life, particularly if mortality is low for adults. Predators consume many turtle eggs on most unprotected beaches. Demographic analyses suggest that the reproductive value of a turtle egg is low and that the sensitivity of population growth to the loss of an egg also is low; sea turtle populations under normal conditions appear to be adapted to withstanding substantial egg loss. However, demographic analyses suggest that the reproductive value of a large juvenile, subadult, or adult sea turtle is higher than that of an egg. Because population growth is most sensitive to changes in survivorship of large juveniles and subadults, we conclude that reduction of human-induced mortality in these life stages will have a significantly greater effect on population growth than reduction of human-induced mortality of eggs and hatchlings.

However, every age and life stage has value. Given that sea turtle species are threatened with extinction, every individual in every life stage becomes important to the survival of the species and protective efforts should be focused on all life stages, even those where individual reproductive values are relatively low.

6
Sea Turtle Mortality Associated with Human Activities

Sea turtles on nesting beaches are most susceptible to mortality associated with human activities at the egg, hatchling, and nesting female stages and in coastal waters at the subadult (including juvenile) and adult stages. They are vulnerable to diverse potentially lethal interactions with human activities, situations including direct predation and habitat modification, incidental capture or entanglement in fishing gear, and physical damage caused by dredging of shipping channels, collisions with ships and boats, and oil-rig removal or other underwater explosions. Each species in the pelagic environment is vulnerable to ingestion of plastics, debris, and petroleum residues. The species differ in behavior and habitat requirements, so they can be affected differently by various human activities.

The recognized sources of mortality related to human activities are listed in order of estimated importance in Table 6-1 for all life stages, and order-of-magnitude mortality estimates are presented in Table 6-2 for juvenile plus adult loggerheads and Kemp's ridleys. The latter table includes the committee's judgment of the certainty of the information on which the estimates were based and lists the preventive and mitigative measures that are in place or being developed. The preventive and mitigative measures are described and evaluated in detail in Chapter 7. The present chapter discusses the information on each mortality factor associ-

TABLE 6-1 A qualitative ranking of the relative importance of various mortality factors on juveniles or adults, eggs, and hatchlings with an indication of mortality caused primarily by human activities. Sources are listed in order of importance to juveniles or adults, because this group includes the life stages with greatest reproductive values.

		Life Stage		
Source of Mortality	Primarily Human Caused	Juveniles to Adults	Eggs	Hatchlings
Shrimp trawling	yes	high	none	unimportant
Other fisheries	yes	medium to low	none	unimportant
Non-human predators	no	low	high	high
Weather	no	low	medium	low
Beach development	yes	low	medium	low
Disease	no	low	unimportant	low
Dredging	yes	low	unimportant	unimportant
Entanglement	yes	low	unimportant	low
Oil-platform removal	yes	low	none	unimportant
Collisions with boats	yes	low	none	unimportant
Directed take	yes	low	medium	unimportant
Power plant entrainment	yes	low	none	unimportant
Recreational fishing	yes	low	none	unimportant
Beach vehicles	yes	low to unimportant	medium	unimportant
Beach lighting	yes	low to unimportant	unimportant	medium
Beach replenishment	yes	unimportant	low	low
Toxins	yes	unknown	unknown	unknown
Ingestion of plastics, debris	yes	unknown	none	unknown

ated with human activities first for eggs and hatchlings and then for juveniles through adults.

The analyses in Chapter 5 on the reproductive value of various life stages called attention to the mortality factors that are most important for juveniles and adults in the ocean and inshore marine habitats. The most important identifiable source of mortality for loggerhead and Kemp's ridleys is incidental capture in shrimp trawls (Table 6-2); other fisheries and fishery-related activities are also important, but collectively only one-tenth as important as shrimp trawling. Dredging, collisions with boats, and oil-rig removal are also important, but only one-hundredth as important as shrimp trawling. Mortality from entrainment in power plants and directed capture of juveniles and adults is believed to be generally low. Parasites,

TABLE 6-2 Order-of-magnitude estimates of human-caused mortality on juvenile to adult loggerhead and Kemp's ridley sea turtles, an index of the certainty of the mortality estimates, and a list of preventive or mitigative measures needed or in place for each type of mortality.

Source of Mortality Caused by Humans	Mortality (number/year)		Rank of Certainty of Estimate*	Preventive and Mitigative Measures in Place
	Loggerheads	Kemp's Ridleys		
Shrimp trawling	5,000-50,000	500-5,000	1	Turtle excluder devices, tow time, time and place restrictions
Other fisheries (trawl and release, passive gear, including entanglement in lost nets and debris)	500-5,000	50-500	3	Open and closed seasons and fisheries, and Marine Pollution International Protocol
Dredging	50-500	5-50	2	Seasons and turtle removal
Collisions with boats	50-500	5-50	3	None
Oil-rig removal	10-100	5-50	3	Surveys and turtle removal
Entrainment in power plants	5-50	5-50	1	Turtle removal with tended barrier nets
Directed take	5-50	5-50	3	Prohibition

*1 = most certain, 3 = least certain.

toxins, and ingestion of plastics and other debris also constitute problems, but present information does not allow quantitative estimates of annual mortality related to them.

MORTALITY OF SEA TURTLE EGGS AND HATCHLINGS

Beach Erosion and Accretion

Erosion of nesting beaches can result in loss of suitable nesting habitat. Erosion rates are influenced by dynamic coastal processes, including sea-level rise. Human interference with natural processes through coastal development and associated activities has resulted in accelerated erosion rates in some localities and interruption of natural shoreline migration. Accretion (deposition of beach sediments) also kills eggs in a nest.

Beach Armoring

Where beach-front development occurs, a site is often fortified to protect the property from erosion. Shoreline engineering is expensive and is virtually always carried out to save structures, not sandy beaches; it usually accelerates beach erosion (NRC, 1987). Several types of shoreline engineering, collectively referred to as beach armoring, include sea walls, rock revetments, riprap, sandbag installations, groins, and jetties. Those structures can cause severe adverse effects on nesting turtles and their eggs. Beach armoring can result in permanent loss of a dry nesting beach through accelerated erosion and prevention of natural beach and dune accretion, and it can prevent or deter nesting females from reaching suitable nesting sites. Clutches deposited seaward of the structures can be inundated at high tide or washed out by increased wave action near the base of them. As the structures fail and break apart, they spread debris on the beach, which can further impede access to suitable nesting sites and result in a higher incidence of false crawls (non-nesting emergences of females) and trapping of hatchlings and nesting turtles. Sandbags are particularly susceptible to rapid failure, which results in extensive debris on nesting beaches. Rock revetments, riprap, and sandbags can cause nesting turtles to abandon nesting attempts or to construct egg cavities of improper size and shape.

Groins are designed to trap sand during transport in longshore currents, and jetties might keep sand from flowing into channels. Those structures prevent normal sand transport and accrete beaches on one side of the structure while starving opposite beaches, thereby causing severe erosion (NRC, 1987) and corresponding degradation of nesting habitat. Even widely spaced groins can deter nesting.

Drift fences, also commonly called sand fences, are erected to build and stabilize dunes by trapping sand that moves along the beach and preventing excessive sand loss. They also protect dune systems by deterring public access. Because of their construction, improperly placed drift fences can impede nesting and trap emergent hatchlings.

Beach Nourishment

Beach nourishment consists of pumping, trucking, or otherwise depositing sand on the beach to replace what has been lost to erosion. Beach nourishment can disturb nesting turtles and even bury turtle nests during the nesting season. The sand brought in might differ from native beach sediments and can affect nest-site selection, digging behavior, incubation temperature (and hence sex ratios), gas-exchange characteristics in incubating nests, moisture content of a nest, hatching success, and hatchling emergence success (Mann, 1977; Ackerman, 1980; Mortimer, 1982b; Raymond, 1984; Nelson, 1986). Beach nourishment can result in severe compaction or concretion of the beach. The trucking of sand to protect beaches can itself increase compaction.

Significant reductions in nesting success on severely compacted beaches have been documented (Raymond, 1984). Nelson and Dickerson (1989a) evaluated compaction on 10 nourished east coast Florida beaches and concluded that five were so compacted that nest digging was inhibited and another three might have been too compacted for optimal digging. They further concluded that, in general, beaches nourished from offshore borrow sites are harder than natural beaches and that, although some might soften over time through erosion and accretion of sand, others can remain hard for 10 years or more. Nourished beaches develop steep escarpments in the midbeach zone that can hamper or prevent access to nesting sites. Nourishment projects involve use of heavy machinery, pipelines, increased human activity, and artificial lighting. They are normally conducted 24 hours a day and can adversely affect nesting and hatching activities. Pipelines and heavy machinery can create barriers to nesting females emerging from the surf and crawling up the beach, and so increase the incidence of false crawls. Increased human activity on a project beach at night might cause further disturbance to nesting females. Artificial lights along a project beach and in the nearshore area of the borrow site might deter nesting females and disorient emergent hatchlings on adjacent nonproject beaches.

Artificial Lighting

Extensive research has demonstrated that emergent hatchlings' principal cues for finding the sea are visual responses to light (Daniel and Smith, 1947; Hendrickson, 1958; Carr and Ogren, 1960; Ehrenfeld and Carr, 1967; Dickerson and Nelson, 1989). Artificial beachfront light from buildings, streetlights, dune crossovers, vehicles, and other sources has been documented in the disorientation of hatchling turtles (McFarlane, 1963; Philibosian, 1976; Mann, 1977; Fletemeyer, 1980; Ehrhart, 1983). The results of disorientation are often fatal. As hatchlings head toward lights or meander along the beach, their exposure to predators and likelihood of desiccation are greatly increased. Disoriented hatchlings can become entrapped in vegetation or debris, and many hatchlings have been found dead on nearby roadways and in parking lots after being struck by vehicles. Hatchlings that find the water might be disoriented after entering the surf zone or while in nearshore water. Intense artificial light can even draw hatchlings back out of the surf (Carr and Ogren, 1960; pers. comm., L. Ehrhart, University of Central Florida, 1989). In 1988, 10,155 disoriented hatchlings were reported to the Florida Department of Natural Resources.

The problem of artificial beachfront lighting is not restricted to hatchlings. Carr et al. (1978), Ehrhart (1979), Mortimer (1982b), and Witherington (1986) found that adult green turtles avoided bright areas on nesting beaches. Raymond (1984) indicated that adult loggerhead emergence patterns were correlated with variations in beachfront light in southern Brevard County, Florida, and that nesting females avoided areas where beachfront light was most intense. Witherington (1986) noted that loggerheads aborted nesting attempts at a greater frequency in lighted areas. Problem lights might not be restricted to those placed directly on or near nesting beaches. The background glow associated with intensive inland light, such as that emanating from nearby large metropolitan areas, can deter nesting females and disorient hatchlings that are navigating the nearshore waters. Cumulatively, along the heavily developed beaches of the southeastern United States, the negative effects of artificial light are profound.

Beach Cleaning

Several methods are used to remove human-caused and natural debris from beaches, including mechanical raking, hand raking, and hand picking of debris. In mechanical raking, heavy machinery can repeatedly traverse nests and potentially compact the sand above them; it also results in tire ruts along the beach that might hinder or trap emergent hatchlings (Hosier et al., 1981). Mann (1978) suggested that mortality within nests can increase when beach-cleaning machinery exerts pressure on soft beaches with large-grain sand. Mechanically pulled rakes and hand rakes can penetrate the surface and disturb a sealed nest or might even uncover pre-emergent hatchlings near the surface of the nest. In some areas, collected debris is buried on the beach; this can lead to excavation and destruction of incubating egg clutches. Disposal of debris near the dune line or on the high beach can cover incubating egg clutches, hinder and entrap emergent hatchlings, and alter nest temperatures. Mechanical beach cleaning is sometimes the sole reason for extensive nest relocation.

Increased Human Presence

Resident and tourist use of developed (and developing) nesting beaches can adversely affect nesting turtles, incubating egg clutches, and hatchlings. The most serious threat caused by increased human presence on the beach is the disturbance of nesting females. Nighttime human activity can cause nesting females to abort nesting attempts at all stages of the process. Murphy (1985) reported that beach disturbance can cause turtles to shift their nesting beaches, delay egg-laying, and select poor nesting sites. Davis and Whiting (1977) reported significantly higher rates of false crawls on nights when tagging patrols were active on an otherwise remote, undeveloped nesting beach. Nesting beaches heavily used by pedestrians might have low rates of hatchling emergence, because of compaction of the sand above nests (Mann, 1977), and pedestrian tracks can interfere with the ability of hatchling loggerheads to reach the ocean (Hosier et al., 1981). Campfires and the use of flashlights on nesting beaches disorient hatchlings and can deter nesting females (Mortimer, 1989).

Recreational Beach Equipment

Recreational material on nesting beaches (e.g., lounge chairs, cabanas, umbrellas, boats, and beach cycles) can deter nesting attempts and interfere with incubating egg clutches and the seaward journey of hatchlings. The documentation of false crawls near such obstacles is increasingly common as more recreational equipment is left in place all night on nesting beaches. There are also reports of nesting females that become entrapped under heavy wooden lounge chairs and cabanas on southern

Florida nesting beaches (pers. comm., S. Bass, Gumbo Limbo Nature Center, 1989; pers. comm., J. Hoover, Dade County Beach Department, 1989). Recreational beach equipment placed directly above incubating egg clutches can hamper emergent hatchlings and can destroy eggs by penetration directly into a nest (pers. comm., C. LeBuff, Caretta Research, Inc., 1989).

Beach Vehicles

The operation of motor vehicles on turtle nesting beaches is still permitted in many areas of Gulf of Mexico and Atlantic states (e.g., Florida, North Carolina, and Texas). Some areas restrict night driving, and others permit it. Driving on beaches at night during the nesting season can disrupt the nesting process and result in aborted nesting attempts. The adverse effect on nesting females in the surf zone can be particularly severe. Headlights can disorient emergent hatchlings and vehicles can strike and kill hatchlings attempting to reach the ocean. The tracks and ruts left by vehicles traversing the beach interfere with the ability of hatchlings to reach the ocean. The time spent in traversing tire tracks and ruts can increase the susceptibility of hatchlings to stress and predation during transit to the ocean (Hosier et al., 1981). Driving directly above incubating egg clutches compacts the sand and can decrease hatching success or kill pre-emergent hatchlings (Mann, 1977). In many areas, beach-vehicle driving is the only reason nests have to be relocated. Vehicular traffic on nesting beaches also contributes to erosion, especially during high tides or on narrow beaches, where driving is concentrated on the high beach and foredune.

Exotic Dune and Beach Vegetation

Non-native vegetation has been intentionally planted in or has invaded many coastal areas and often displaces native species, such as sea oats, beach morning glory, railroad vine, sea grape, dune panic grass, and pennywort. The invasion of such destabilizing vegetation can lead to increased erosion and degradation of suitable nesting habitat. Exotic vegetation can also form impenetrable root mats, which can prevent proper nest-cavity excavation, and roots can penetrate eggs, cause eggs to desiccate, or trap hatchlings.

The Australian pine (*Casuarina equisetifolia*) is particularly detrimental. Dense stands of that species have taken over many coastal strand areas throughout central and southern Florida, causing excessive shading of the beach. Studies in southwestern Florida suggest that nests laid in the shaded areas are subjected to lower incubation temperatures, which can alter the natural hatchling sex ratio (Marcus and Maley, 1987; Schmelz and Mezich, 1988). Fallen Australian pines limit access to suitable nest sites and can entrap nesting females. Davis and Whiting (1977) reported that nesting activity declined in Everglades National Park where dense stands of Australian pine took over native beach vegetation. Schmelz and Mezich (1988) indicated that dense stands of Australian pines in southwestern Florida affect nest-site selection and cause increased nesting in the middle beach area and higher ratios of false crawls to nests compared with areas of native vegetation.

MORTALITY OF SEA TURTLE JUVENILES AND ADULTS

Shrimp Fishing

Description of the Fishery

The shrimp fishery has the highest product value of any fishery in the United States. It also is the most important human-associated source of deaths of adult and subadult sea turtles. Sea turtles are captured in shrimp trawls towed along the bottom behind shrimping vessels. The vessels might tow one to four otter trawls. An otter trawl consists of a heavy mesh bag with tapered wings on each side that funnel shrimp into the cod end, or bag, of the net. To keep the trawl near the bottom and achieve horizonal opening of the mouth of the trawl, a weighted otter board is positioned at the front of each wing to serve as a hydrofoil. Turtles swimming, resting, or feeding on or near the bottom in the path of a trawl are overtaken and enter the trawl with the shrimp.

What is often perceived as the U.S. shrimp fishery is actually a number of fisheries. Seven species of shrimp are harvested in the fishery: brown shrimp (*Penaeus aztecus*), white shrimp (*P. setiferus*), pink shrimp (*P. duorarum*), seabobs (*Xiphopenaeus kroyeri*), royal red shrimp (*Hymen openaeus robustus*), rock shrimp (*Sicyonia brevirostris*), and trachs (*Trachypenaeus* sp.). Each shrimp species is taken by a distinct fishery, and the several fisheries are differentiated according to fishing depths, seasonal landings, vessel and gear, fishing localities, fishing techniques, and other characteristics.

The most valuable shrimp species in the United States are brown, white, and pink. For example, in 1985, U.S. commercial shrimp catches were 122,000 metric tons in the gulf and 13,000 metric tons in the south Atlantic. The white shrimp fishery is the most important in the U.S. south Atlantic; the brown shrimp fishery is more important in the gulf.

Brown shrimp range along the north Atlantic and Gulf of Mexico coasts from Martha's Vineyard, Massachusetts, to the northwestern coast of Yucatan. The range is not continuous, but is marked by an apparent absence of brown shrimp along Florida's west coast between the Sanibel and the Apalachicola shrimping grounds (Farfante, 1969). In the U.S. Gulf of Mexico, catches are highest along the coasts of Texas, Louisiana, and Mississippi. Brown shrimp can be caught at depths of 100 m or more, but most come from depths less than 50 m. The season begins in May, peaks in June and July, and declines to an April low (Gulf of Mexico Fishery Management Council, 1981).

White shrimp range along the Atlantic coast from Fire Island, New York, to Saint Lucie Inlet, Florida, and along the gulf coast from the mouth of the Ochlockonee River, in the Florida panhandle, to Campeche, Mexico. In the gulf, there are two centers of abundance: one along the Louisiana coast and one in the Campeche area. White shrimp are comparatively shallow-water shrimp; most of the catch comes from depths less than 25 m. The catch has a major peak in late summer and early fall, with an October high and a minor peak of over-winter shrimp with a peak in May. The largest catches occur west of the Mississippi River to the Freeport, Texas, area, although the catch is considerable along the entire north central and western gulf and south Atlantic. Pink shrimp range along the Atlantic from the lower Chesapeake Bay to the Florida Keys and around the gulf coast to the Yucatan peninsula. Major concentrations exist off southwestern Florida and in the southeastern part of the Gulf of Campeche. The two major pink shrimp grounds in the United States are the Tortugas and Sanibel grounds in southwestern Florida. The pink shrimp catch comes mainly from depths less than 50 m, with a maximal catch from 20-25 m. Most of the catch is taken off Florida and is greatest in the southwestern waters of the state. The catch is high from October through May.

In the south Atlantic, white shrimp account for the majority of landings in Georgia and the Atlantic coast of Florida. In South Carolina, small landings of white shrimp in the spring are augmented by a much larger catch in the fall. The spring white shrimp fishery is based on adults that have over-wintered, whereas the fall catch is based almost entirely on young of the year. White shrimp are caught in North Carolina principally during the fall, but the catch is much smaller than that of brown and pink shrimp (Calder et al., 1974). Brown shrimp predominate in the North Carolina fishery. During some years, catches of brown shrimp exceed those of white shrimp in South Carolina as well. The peak of the brown shrimp harvest occurs during the summer in all four south Atlantic states. Brown shrimp enter and leave the Florida east coast fishery earlier than in the other three states. In the south Atlantic, pink shrimp are of major commercial significance only in North Carolina, where they account for about one fourth of the total shrimp landings. Fishing for pink shrimp usually begins in the spring and ends by midsummer.

Other minor shrimp species are often fished incidentally or during the offseasons of the major shrimp fisheries. A targeted rock shrimp fishery exists in the south Atlantic off northern Florida from August to January. In recent years, vessels in the western gulf have focused their effort on "trachs" during the late winter and spring months. The trach catch is primarily from depths of 20-50 m. The royal red shrimp fishery is relatively insignificant, occurring at depths of 250-550 m; harvesting and marketing obstructions have limited this fishery. Seabobs are caught most often in shallow waters at 13 m or less and in the open ocean; along the Louisiana coast, catch rates are highest in October-December.

The various fisheries share some similarities, in socioeconomic makeup and biology, but there are important contrasts, principally in depth of operation. The similarities and differences might have an important bearing on turtle bycatch. Many of the fisheries for shrimp, especially of the three major species, are timed and located in relation to the life histories of the shrimp. For example, several discrete fisheries constitute the "gulf brown shrimp fishery." Juvenile and subadult brown shrimp live in bays and estuaries and are harvested by the inshore fishery. The shrimping vessels used are usually small, from 6 to 30 m long; most are about 15 m long.

As the shrimp mature, they migrate offshore. Vessels fish near shore out to a depth of 25 m, especially for subadult and adult white and pink shrimp. The larger vessels of the gulf type begin almost exclusive harvest of the species (adult brown and pink shrimp) in water deeper than 25 m; these vessels are generally 20-30 m long. As the maturing brown shrimp continue to migrate into the deeper gulf waters, the smaller inshore vessels are limited, and only the larger vessels can gain access to the fishery. The offshore fishery provides the basis for the adult brown shrimp fishery.

The pink shrimp fleet off Florida uses a variety of vessels of different sizes but is associated primarily with larger offshore boats. White and brown shrimp are caught in bays and estuaries in some states by the smaller inshore vessels. Unlike the adult brown shrimp fleet, which uses larger vessels, the adult white shrimp fleet uses vessels of all sizes.

Distribution and Intensity of the Fishery

The distribution and intensity of fishing in inside waters were calculated from raw data and summaries provided to the committee by NMFS. For waters outside the coast, the information was taken from Appendix F.

Fishing effort is measured in effort-days (24-hr days of towing time) per boat, regardless of variations in vessel size, the number and size of nets it tows, and water depth. That probably underestimates effort outside the coastal beaches, compared with bays, rivers, and estuaries, because offshore boats tend to be larger and tow more nets for longer periods than the inside boats. The unit of effort is used by NMFS in the Gulf of Mexico and the Atlantic Ocean and is estimated annually with the cooperation of individual state agencies. Quarterly effort in offshore waters is plotted by fishing zone from Texas to Maine for 1987 and 1988 (Figure 6-1). Data used are the best available currently from NMFS.

The shrimp fishery is intense, totalling about 373,000 24-hr days per year from the Mexican border in the gulf to Cape Hatteras in the Atlantic during 1987 and 1988. The intensity is much greater in the gulf—345,000 days—compared with the Atlantic's 28,000 days; 92% of the total effort is expended in the gulf. Most effort is offshore of the coastal beaches, about 249,000 days, or 67% (67% for the gulf and 68% for the Atlantic). The rest is expended in the bays, estuaries, and rivers, 33% of the total effort. The most intense fishery outside the coastal beaches is off Texas and Louisiana and includes 83% of the effort off the coastal beaches and 55% of the total effort from Maine to the Mexican border. Shrimp fishing off Mexico near the U.S. border has been low or absent in recent years.

Although fishing efforts in the bays, estuaries, and rivers are similar in the Atlantic and Gulf of Mexico, distinct differences occur in shrimping efforts directed toward offshore waters. Brown and pink shrimp are important fisheries in the Gulf; therefore, more shrimping effort is expended in deeper waters of the gulf than in the Atlantic, where white shrimp dominate the fishery. Statistical reporting procedures vary between the Atlantic and gulf data bases (pers. comm., J. Nance, NMFS, 1989). Areas of effort are reported by distance from shore in the Atlantic and by depth in the gulf. Because of differences associated with the slope of the gulf's continental shelf, a comparison of effort by distance from shore would be impractical; however, because white shrimp is the principal Atlantic fishery, effort focuses on a relative shallower and nearshore fishery.

For 1987 and 1988, NMFS data indicate that 92% of the Atlantic effort outside of coastal beaches was from 0 to 5 km, 3% was from 5 to 20 km, and 4% was farther than 20 km offshore. In contrast, in the gulf outside of coastal beaches, 65% of the effort was in water shallower than 27 m (a depth contour that ranges from approximately 14-50 km offshore), while 24% was between 27 m and 48 m; another 11% was deeper than 48 m.

Seasonally, effort for the fisheries outside of coastal beaches is greatest in summer and fall, lower in spring, and least in winter. In 1987 and

FIGURE 6-1 Shrimp-fishing effort, 1987-1988, by season. Fishing zones are shown on horizontal axis (see Figure 4-1). Data from Appendix F.

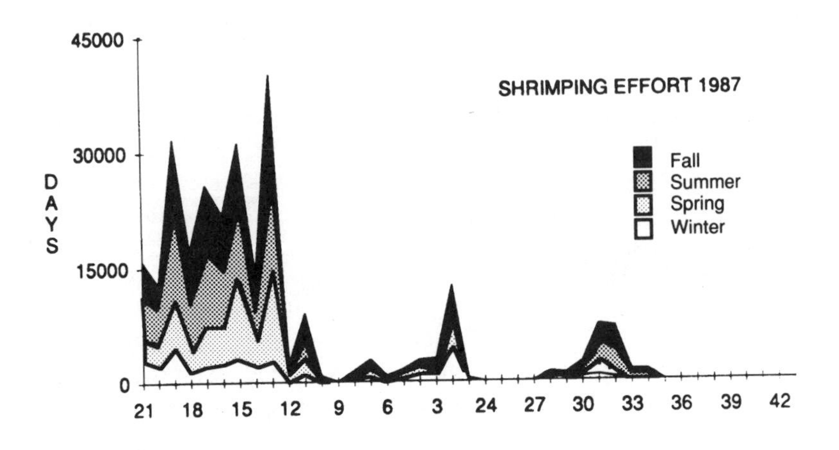

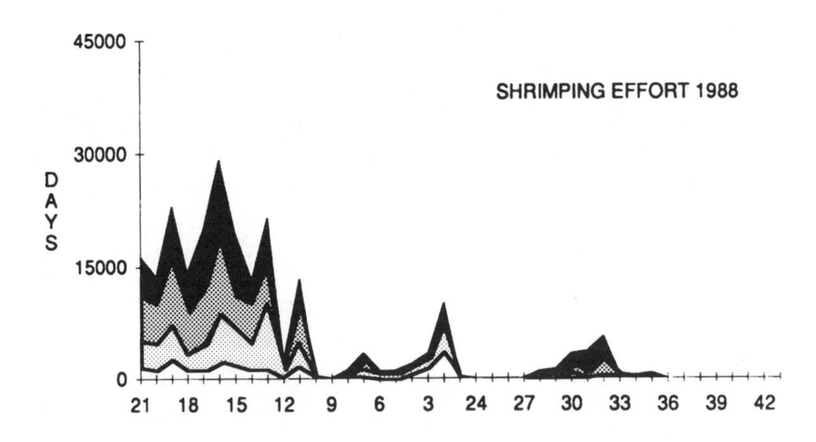

1988, 33% of the effort was in summer, 31% in fall, 24% in spring, and 12% in winter. This pattern largely represents that of the western gulf; local variations from this pattern occur. Off Georgia and the Carolinas, little fishing takes place in winter, whereas off the Atlantic coast of northern Florida, effort is more uniform through the year and includes significant winter fishing. Along the gulf coast of Florida, fishing is most intense in winter and spring.

Fishing effort in the gulf has grown steadily since 1960. The increase has been by a factor of about 2.5 in 30 years. The proportion of the effort in rivers, estuaries, and bays has remained about the same during this growth period. In the Atlantic, comparable data were not available, but from 1984 to 1988, total effort ranged from 24,000 to 34,000 24-hr days per year, reaching a maximum in 1986.

Seasonal Changes in Stranding, Shrimp-Fishing Effort, and Turtle Abundance

The recent abundance of stranded sea turtles and the intensity of shrimp fishing vary from the western Gulf of Mexico along the coast to the Gulf of Maine and from season to season. The distributions of strandings are complex interactions between trawling intensity and the abundance of sea turtles and other factors. The relationship between stranding and fishing intensity takes on a different perspective when viewed on short and long time and space scales. For example, the highest stranding rate does not occur off Texas and Louisiana (Figure 4-3), where shrimp fishing is now most intense (Figure 6-1); turtle abundance is lower there now than along the south Atlantic coast. Such broad-scale comparisons do not provide evidence of the present effects of trawling, because they do not account for historical changes in the abundance of turtles in relation to past shrimping and other mortality factors.

The relation between turtle stranding and fishing effort on an intermediate scale—i.e., seasonal changes in areas that differ in the ratio of turtle abundance to shrimping effort—permits an interesting, but speculative interpretation. Sites chosen for our analysis were those with NMFS contractual stranding surveys: Texas (zones 17-21), the gulf coast of Florida (zones 4 and 5), the northern Florida's Atlantic coast (zones 29-31), and Georgia-South Carolina (zones 31 and 32). Those four areas span a range of ratios of turtle abundance in aerial surveys (number of turtles sighted/10,000 km^2) to shrimp fishing effort (10,000 24-hr days of fishing) from about two for Texas to about 2,500 off Florida, or a factor of about 1,250 in the ratio of abundance of turtles to fishing effort in the two states.

In only two of the eight examples (Figure 6-2) was turtle stranding positively correlated with fishing effort ($p = 0.05$). One of those examples, from the Atlantic coast of Florida in 1988, was used by Schroeder and Maly (1989) as evidence for a direct relation between stranding and fishing effort. The relation between stranding and effort is more complex than the simple argument that more shrimping effort equals more turtle stranding.

Models of fishing-induced mortality have produced insights that can be applied to the present situation. Results of an examination of the season-

FIGURE 6-2 Seasonal changes in sea-turtle strandings on ocean beaches and shrimp-fishing effort offshore of ocean beaches at four locations along the Gulf of Mexico and the Atlantic coast for 1987 and 1988. The four areas differ greatly in the abundance of turtles sighted from aerial ocean surveys and shrimp-fishing effort. Texas (zones 18-21) had the fewest turtles per unit of shrimp fishing, followed by western Florida (zones 4-5), and Georgia and South Carolina (zones 31-32); the largest number of turtles per unit of shrimp-fishing effort was for Atlantic north Florida (zones 28-30). The correlations and p values are those for a simple linear regression. Data from Appendixes E and F.

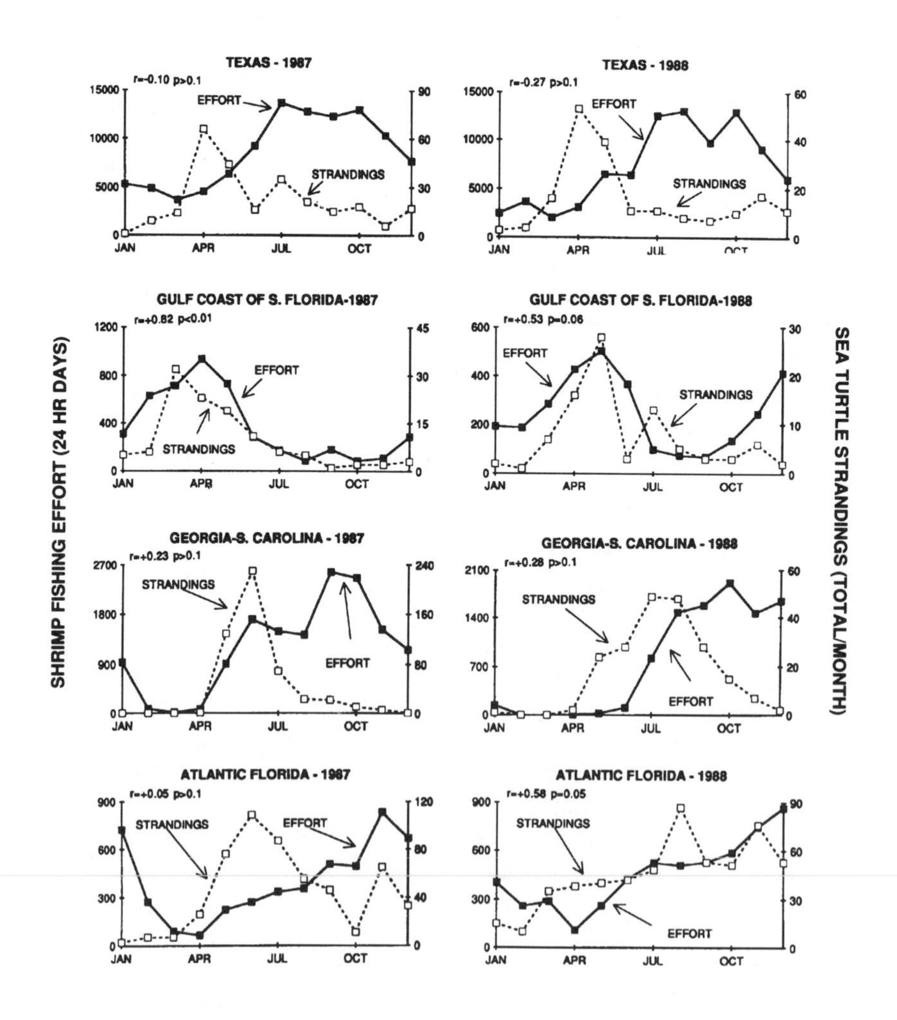

al changes in fishing effort and stranding in each area and year (Figure 6-2) suggest an analogy with those models. In Texas, for example, stranding reached a maximum in April and then declined as effort increased; later in the summer, effort was high, but few turtles were stranded. One possible interpretation is that trawling has eliminated most of the turtles in that area by early summer. Alternative explanations could be that the turtles migrate through the area in the spring (Pritchard and Márquez M., 1973; Timko and Kolz, 1982) and that oceanic conditions in the spring differ from those in the other seasons and tend to bring more dead floating turtles to the beach than in other seasons (Amos, 1989).

A pattern with some similar features was observed on the gulf coast of southern Florida. The decline in strandings occurred while effort was high, as would be expected from fishing-induced mortality. Effort declined, but the decline in turtle stranding began before effort dropped to low levels.

In the Georgia-South Carolina region, stranding was greatest as the fishing effort was increasing early in the season, but later declined as effort continued to increase. That pattern is consistent with the interpretation that fishing effort locally depleted the turtles by middle to late summer.

Finally, on the Atlantic coast of northern Florida, stranding reached a maximum as effort increased and then began to decline—sharply in 1987, marginally in 1988. That pattern is also consistent with the effects of fishing-induced mortality on a fixed or limited population size. The only case of no major decline when effort was high was the northern Florida example of Schroeder and Maley (1989). That area has the most turtles per unit of shrimping effort among the four locations examined and would be the most likely to support a direct relation between stranding and effort over an extended period with modest levels of fishing effort relative to the standing stock of sea turtles.

We cannot eliminate the alternative hypotheses from the existing data on turtle migration and ocean currents. However, these observations are consistent with models of fishing-induced mortality, and that suggests that this is a likely hypothesis. It might explain the lack of a significant positive correlation between seasonal fishing effort and turtle stranding in all but two of the eight examples: the relationship between fishing effort and abundance of the fished species often are out of phase.

We note that neither significant positive nor nonsignificant negative correlations between seasonal changes in stranding and shrimping effort are by themselves enough to reveal the influence of shrimping on stranding. The relationships are more complex on these broad temporal and spatial scales in response both to shrimping effort and to changes in turtle abundance. The influence of shrimping on turtles cannot be excised from the seasonal patterns only by a simple linear regression analysis. More incisive analyses, as presented below, are needed to tease apart the relationship.

Strong Evidence of Shrimp Trawling as an Agent of Sea Turtle Mortality

One central charge of this committee is to evaluate available evidence to assess whether incidental catch of sea turtles during shrimp trawling is indeed a cause of sea turtle mortality and, if so, to estimate the magnitude and importance of this mortality. Sea turtles are undoubtedly caught in large numbers during shrimp trawling. For example, the primary source of tag returns from female Kemp's ridleys tagged at the nesting beach at Rancho Nuevo (84% of 129 returns) has come from incidental capture of the turtles and reporting of tag numbers by cooperative shrimpers (Pritchard and Márquez M., 1973; Márquez M. et al., 1989). Furthermore, observers on vessels conducting commercial shrimp trawling have reported large numbers of sea turtle captures (Hillestad et al., 1978; Roithmayr and Henwood, 1982).

Even if individual fishermen catch few turtles, the size of the shrimp fleet and the effort exerted result in a collective catch that is "large," although not all sea turtles that are caught in shrimp trawls necessarily die as a result. In a recent review, 83% of 78 papers on the incidental capture of all Atlantic sea turtle species in fishing operations inferred that shrimp trawling is a major source of mortality (Murphy and Hopkins-Murphy, 1989).

We consider below five observations that, when taken together, constitute a compelling demonstration that incidental capture during shrimp trawling is the proximate cause of mortality of substantial numbers of sea turtles.

Relation Between Sea Turtle Mortality in Trawls and Tow Time The most convincing data available to assess whether shrimp trawling is responsible for sea turtle deaths come from NMFS studies relating the time that a trawl was allowed to fish (tow time) to the percentage of dead sea turtles among those captured. Henwood and Stuntz (1987) published a linear equation showing a strong positive relation between tow time and incidence of sea turtle death. They concluded that "the dependence of mortality on tow time is strongly statistically significant ($r = 0.98$, $p < 0.001$)."

The committee analyzed the data set used by Henwood and Stuntz to clarify in detail the relationship between tow times and mortality. Death rates are near zero until tow times exceed 60 minutes; then they rise rapidly with increasing tow times to around 50% for tow times in excess of 200 minutes. That pattern is exactly what would be expected if trawling were causing the drowning of an air-breathing animal. Death rates never reach 100%, because some turtles might be caught within 40-60 minutes of lifting the net from the water. The data provide the functional relation between other correlative relations, namely, between fishing activity and dead turtles or population trends.

Under conditions of involuntary or forced submergence, as in a shrimp trawl, sea turtles maintain a high level of energy consumption, which rapidly depletes their oxygen store and can result in large, potentially harmful internal changes. Those changes include a substantial increase in blood carbon dioxide, increases in epinephrine and other hormones associated with stress, and severe metabolic acidosis caused by high lactic acid concentrations. In forced submergence, a turtle becomes exhausted and then comatose; it will die if submergence continues. Physical and biological factors that increase energy consumption, such as high water temperature and increased metabolic rates characteristic of small turtles, would be expected to exacerbate the harmful effects of forced submergence because of trawl capture.

Drowning can be defined as death by asphyxiation because of submergence in water. There are two general types of drowning: "dry" and "wet." In dry drowning, the larynx is closed by a reflex spasm, water is prevented from entering the lungs, and death is due to simple asphyxiation. In wet drowning, water enters the lungs. For nearly drowned turtles, the wet type would be more serious, because recovery could be greatly compromised by lung damage due to inspired seawater. The exact mechanism of sea turtle drowning is not known, but a diagnostic condition of the wet-drowning syndrome—the exudation of copious amounts of white or pink froth from the mouth or nostrils—has been observed in trawl-captured turtles.

Turtles captured in shrimp trawls might be classified as alive and lively, comatose or unconscious, or dead. A comatose turtle looks dead, having lost or suppressed reflexes and showing no sign of breathing for up to an hour. The heart rate of such a turtle might be as low as one beat per 3 minutes. Lactic acid can be as high as 40 mM, with return to normal values taking as long as 24 hours. It takes 3-5 hours for lactic acid to return to 16-53% of peak values induced by trawl capture. Although the fate of comatose turtles directly returned to the sea is unknown, it is reasonable to assume that they will die (Kemmerer, 1989).

In 1989, NMFS conducted a tow-time workshop to analyze data on tow times and turtle conditions from seven research projects. The projects spanned 12 years, during which 4,397 turtles were encountered. The numbers of dead and comatose turtles increased with tow time (Figure 6-

3). Small increases in tow time between 45 and 125 minutes resulted in large, steep increases in the numbers of dead and comatose turtles. For most tow times, there were more comatose than dead turtles. Few turtle deaths were related to tow times of less than 60 minutes. Tow times are thus a critical element in determining turtle mortality associated with shrimp trawls.

Coincidence of Opening and Closing of Shrimp Season with Changes in Turtle Stranding on Adjacent Beaches in Texas and South Carolina Murphy and Hopkins-Murphy (1989) used the data on sea turtle stranding in South Carolina in 1980-1986 to seek a temporal relation between the opening of the ocean shrimp fishery and the rate of stranding. In South Carolina, the Sea Turtle Stranding and Salvage Network (STSSN) has provided complete and reliable coverage of the ocean beaches for several years. The opening of the ocean shrimp fishery took place between May 16 and June 26 and varied from year to year. The 7-year total number of strandings (190 carcasses) in the 2-week periods just after the opening of the fishery was 5 times as large as the number of strandings in the 2-week periods immediately before the opening (38 carcasses). Although that does not conclusively demonstrate a causal relationship, repetition of the

FIGURE 6-3 Relation between the percentage of dead or dead and comatose loggerheads as a function of tow time of trawls. Total number of turtles captured was 4,397. Compiled by the committee from raw data provided by NMFS that were the basis for Henwood and Stuntz's (1987) calculations.

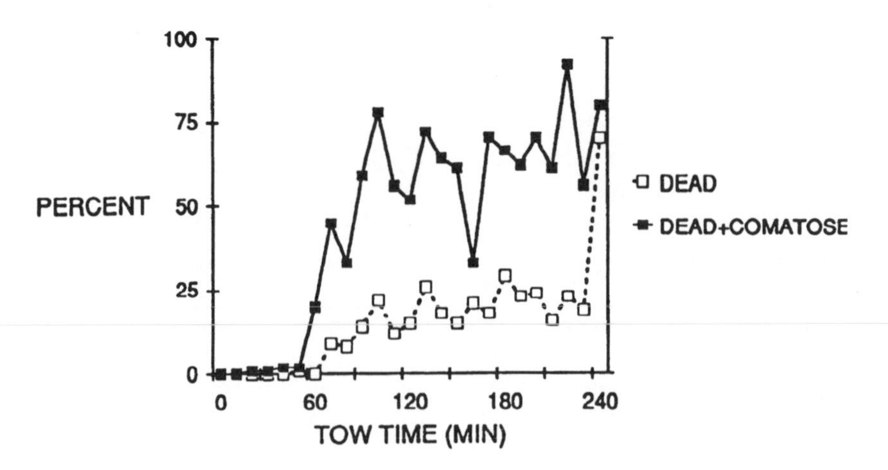

TABLE 6-3 Changes in the number of stranded sea turtles before and after the opening and closing of shrimp fishing seasons using data from STSSN.

Examples	Years	Mean Number of Strandings*				Significance of Differences Probability Level†		
		2-4 Weeks Before	0-2 Weeks Before	0-2 Weeks After	2-4 Weeks After			
		a	b	c	d	a to b	b to c	c to d
South Carolina Opening	1980-1989	8	10	37	34	0.48	0.006	0.96
Texas Opening	1980-1988	1	1	6	4	0.58	0.04	0.36
Texas Closure	1980-1988	10	8	3	1	0.56	0.03	0.008

*Loggerheads or, for South Carolina, mostly loggerheads
†Two-tailed Wilcoxon matched-pair, signed rank test

FIGURE 6-4 Sea turtle strandings on beaches before and after opening or closing of shrimping seasons in South Carolina and Texas. Statistical analysis of differences is in Table 6-3. (Compiled from NMFS data.)

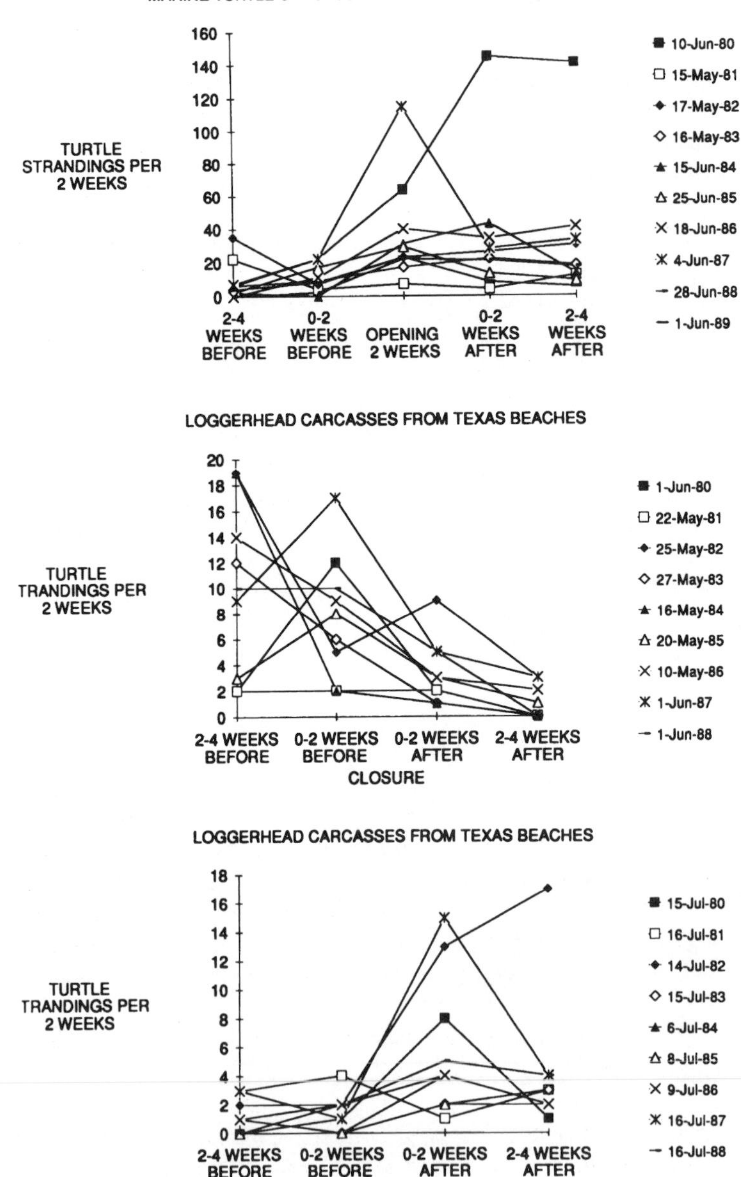

large increase in stranding after the beginning of shrimping, despite variation in the date of the beginning of shrimping, strongly suggests that shrimp trawling is the proximate cause of the large increase in dead sea turtles found on South Carolina beaches after the opening of shrimp season.

To evaluate further the potential effect of shrimp trawling on the numbers of sea turtles found dead on South Carolina beaches, we followed the lead of Murphy and Hopkins-Murphy (1989) and segregated stranding data into two-week intervals (the first and second halves of each month, because of how the data were compiled) for the 10-year STSSN data base (1980-1989). The 2-week interval in which the fishery opened was designated as the "2 weeks after opening," unless the opening occurred at the end of the 2 weeks, in which case the next 2-week interval was called the "2 weeks after opening." We compiled the strandings not only for the 2-week intervals before and after opening of the shrimp season, but also for the 2-week periods before and after that 4-week period, for a total of four 2-week periods (Table 6-3, Figure 6-4). We then used the Wilcoxon signed-ranks test (a paired-sample nonparametric test) to compare strandings in each pair of successive 2-week periods. The 3.7-fold increase in turtle strandings that occurred in the 2 weeks after opening has a two-tailed probability of 0.006 of occurring by chance. No other contrast between successive 2-week intervals had a probability of less than 0.10. This analysis thus implies that shrimp trawling was indeed responsible for the increase in turtle strandings. It is an especially strong analysis, in that the increase observed with the opening of the fishery was independent of seasonal changes (the date of opening varied widely—from May 16 to June 26).

We also used the STSSN data base for Texas beaches for the 9 years of 1980-1988 to evaluate the effects of fishery closing and opening on stranding of loggerheads (Table 6-3, Figure 6-4). The changes in four consecutive 2-week periods were compared and analyzed as for the South Carolina data. The application of the nonparametric tests demonstrated that the sixfold increase in loggerhead stranding between the 2 weeks before and the 2 weeks after opening of the Texas brown shrimp fishery had a two-tailed probability of 0.04 (Table 6-3). Differences between 2-4 weeks before and 0-2 weeks before intervals were not statistically significant. As in the South Carolina case, the statistical tests suggest that loggerhead stranding increased significantly when shrimp trawling opened in Texas.

Finally, we analyzed in the same manner how stranding rates changed at the time of closing of the Texas brown shrimp fishery (Table 6-3, Figure 6-4). Loggerhead stranding decreased between the 2 weeks before and the 2 weeks after closing by a factor of 2.7. The probability that that decrease occurred by chance with a 2-tailed test was 0.01 (Table 6-3). The contrast between the first two periods (2-4 weeks before versus 0-2 weeks before closing) had a probability of 0.56. A decline in stranding did occur between 0-2 weeks after and 2-4 weeks after closing, $p=0.008$. Consequently, although a large and statistically significant decline in loggerhead stranding had occurred after closing of the Texas brown shrimp fishery, the decline continued to occur between the last two periods. Given the uncertainty as to how long it takes for dead turtles to reach the beach, those results are consistent with either an effect of brown shrimping on sea turtle stranding or a general decline in sea turtle stranding during the period for other reasons. Because the dates of closure varied from May 10 to June 1, we interpret the decline to be fishery related.

Stranding in the three cases—South Carolina opening, Texas opening, and Texas closing—changed by factors of 3.9, 5.0, and 4.5, based on the 4 weeks before and 4 weeks after opening. We conclude that, in those locations and at those times, approximately 70-80% of the stranded turtles were caught in shrimp trawls. Taken along with the results on tow time given above, these results provide strong evidence of the crucial role of shrimp fishing on turtle mortality.

Relation Between Loggerhead Populations and Shrimping Effort Along the southeastern Atlantic coast, loggerhead populations are declining where shrimp fishing is intense off the nesting beaches. They are not declining, however, where shrimping effort is low or absent. Nesting populations in South Carolina and Georgia (Figure 3-1*c,d*) are declining, whereas those in central and southern Florida are not and might even be increasing (Figure 3-1*e,f*). Shrimping effort declines markedly to the south at Cape Canaveral (Figure 6-1) so, for example, the population at Hutchinson Island is subject to essentially no shrimp fishing off the nesting beach (fewer than 17 effort-days per year) whereas the populations at Little Cumberland Island, Georgia, and Cape Island, South Carolina, have intense fisheries (about 400-7,000 days per year per fishing zone). Further evidence of the relation between shrimping effort and turtle population declines is found in the lower stranding rates of loggerheads in fishing zones 26-28 at Canaveral and south, where effort is low, even though these zones have the highest density of nesting loggerheads (Figure 4-2). Shrimping effort declines from about 1,000 effort-days per year in zone 28 to almost none in zones 27 and 26. In contrast, effort increases irregularly north of zone 28 to a maximum of 5,000-7,000 in zone 32, off South Carolina. Because the turtles aggregate off the nesting beaches during the nesting seasons between their multiple nestings, the absence of the fishery would be expected to reduce mortality and contribute to the maintenance or growth of local nesting populations, as was observed south but not north of Canaveral.

Quantification of Sea-Turtle Mortality in Shrimp Trawls Incidental catch and mortality of sea turtles in shrimp trawls have been estimated on the basis of interviews with vessel captains (Anon., 1976; Anon., 1977; Cox and Mauerman, 1976; Rabalais and Rabalais, 1980; Rayburn, 1986) and direct observation by fishery observers on commercial shrimping vessels (Hillestad et al., 1978; Ulrich, 1978; Roithmayr and Henwood, 1982; Henwood and Stuntz, 1987). Henwood and Stuntz (1987) provide the most complete assessment of sea turtle capture and mortality for the south Atlantic and Gulf of Mexico shrimp fisheries. Their study, based on more than 27,000 hours of observed trawling, estimated an annual incidental capture of approximately 47,000 sea turtles, with an estimated mortality of about 11,000. The study suffers from the a posteriori approach of estimating capture and mortality from programs not specifically designed for that purpose and, therefore, is limited in its ability to account for possible differences in capture and mortality related to such variables as species, season, depth, and geographic location. Although the statistics have been debated (Clement Assoc., 1989; Murphy and Hopkins-Murphy, 1989), the estimates are conservative because of the approach taken. Points of contention with the estimates of mortality include the use of data from a research study of the use of turtle excluders on trawlers, the representativeness of fishing distribution between research studies and commercial shrimping, the precision of mortality estimates based on the method used to calculate mortality rate, the magnitude of mortality estimates based on the assumption that all comatose sea turtles survive, and the magnitude of mortality estimates based on the complete omission of inside waters (waters landward of the barrier islands, including bays, sounds, etc.) (Table 6-4).

The objective of the trawler excluder study was to design and use an apparatus that would effectively prevent the incidental capture of sea turtles in existing shrimping gear. Shrimp fishermen fished with commercial fleets in both the Gulf of Mexico and the south Atlantic. Sixty-two percent of the trips were in the south Atlantic, where 95% of the loggerheads and 78% of the Kemp's ridleys were caught (Table 6-5). Georgia, fishing zone 31, accounted for 74% of the total south Atlantic trips and 58% of the catch of loggerheads and 71% of the catch of Kemp's ridleys in the south Atlantic. For the south Atlantic, the estimated catch rate for the trawler excluder study was strongly influenced by the catch rate off Georgia; the Georgia catch rate was lower than the other zones sampled. Similarly, in the gulf, the catch rates reflected activity off Texas and Louisiana, which comprised 75% of the effort. Eliminated from consideration in this study

TABLE 6-4 Points of contention and potential sources of bias to estimated mortality as calculated by Henwood and Stuntz (1987).

Contention/Potential Bias	Impacts on Estimate
Use of trawl excluder study provided a biased sample because fishermen fished where turtles were.	Fishermen fished with fleet and were not controlled by contracting agency. A sea turtle "hot spot" (Cape Canaveral channel) (see Figure 4-1) was eliminated from study. Georgia (fishing zone 31 (see Figure 4-1)) accounted for the majority of the study effort and catch in south Atlantic; Texas and Louisiana (fishing zones 15-19 (see Figure 4-1)) accounted for the majority of the study effort in the Gulf of Mexico. The study was used to calculate catch rates for the south Atlantic and the gulf. Overall catch (and hence, mortality) was estimated by multiplying catch rates by commercial fishing effort as determined separately by NMFS for the gulf and south Atlantic. No significant bias was detected.
Fishing effort in study did not reflect true commercial fishery. Data were biased.	In the Gulf of Mexico, 65% of the commercial effort was exerted in waters ≤27 m (1988), whereas 84% of the effort reported in the study was in waters ≤27 m. If catch rates are partitioned by depth (≤27 m and >27 m), based on sea turtle distribution (see Chapter 4), the 19% oversampling of water ≤27 m results in an overestimate of catch (and mortality) of about 24% for the gulf.
Precision of mortality estimates is erroneous. Methods used did not incorporate variability of mortality rate into variability of estimated mortality. (Product estimated captures times mortality rate.)	Reported limits of confidence intervals would be widened, thus increasing the uncertainty about the estimated mortality.
All comatose sea turtles were assumed to survive. This produces an underestimate, because not all comatose sea turtles do survive.	If all comatose turtles died, the estimate of mortality would increase from about 11,000 to about 32,000.
Captures in inside waters were not included, thus reported estimates are low.	Reported mortality estimates are underestimates due to omission of inside waters data. Approximately 37% of total shrimping effort occurs in inside waters. Depending on species, total estimated mortality might be higher by a factor of 1.6 or from 11,000 to 18,000.

TABLE 6-5 Distribution of effort (number of tows) and capture of loggerhead and Kemp's ridley sea turtles from trawler excluder study.

Statistical Zone	Tows (no.)	Loggerheads (no.)	Kemp's Ridleys (no.)
1	93	3	
2	476	3	
3	60	2	
15	160	0	
16	111	0	1
17	110	1	
18	1,340	5	2
19	169	1	1
Total Gulf of Mexico	2,519	14	4
30	308	50	
31	3,024	161	10
32	527	41	4
33	209	23	
Total Atlantic	4,068	275	14

Source: Partial data set from Henwood and Stuntz (1987) and W. Stuntz (pers. comm.)

were the catch and effort data from the Cape Canaveral ship channel and surrounding area (approximately 24 km). This local area harbors large concentrations of sea turtles throughout the year, and high turtle catch rates there do not reflect those occurring outside the Canaveral area (Henwood and Stuntz, 1987). Elimination of those data provided conservative estimates of catch rates for the south Atlantic.

Distribution of effort by depth in the Gulf of Mexico in the Henwood and Stuntz (1987) study is biased toward shallower waters than are usual or typical for the commercial shrimp fleet. The commercial fleet exerted 65% of total offshore shrimping effort in 27 m or less in 1988 (pers. comm., E. Klima, NMFS, 1989), whereas 84% of the total effort reported by Henwood and Stuntz (1987) was in 27 m or less. Statistically significant differences in capture rates among depths were not found in these data, and the data were pooled to provide the best estimates of capture rates in the Gulf of Mexico. However, information discussed in Chapter 4 strongly suggests that turtle abundance is negatively correlated with depth.

The confidence intervals associated with estimates of mortality in Henwood and Stuntz (1987) do not incorporate the uncertainties associated with the estimated mortality rate, so they portray a lower level of uncertainty than is reflected by the data. Incorporating that uncertainty would broaden the confidence intervals about the estimates of mortality.

Henwood and Stuntz (1987) restricted their analysis of mortality rate (number of dead turtles per unit of tow time) to turtles classified as dead; they excluded turtles classified as comatose. Recent work (pers. comm., P. Lutz, University of Miami, 1989; Stuntz and Kemmerer, 1989) indicates that some comatose sea turtles die even after proper resuscitation techniques have been applied and the turtle becomes active. Internal injuries that are not visible in turtles landing on deck and are not initially totally debilitating are considered a factor in delayed mortality of trawl-caught sea turtles (pers. comm., D. Owens, Texas A&M University, 1989). If some or all comatose sea turtles die as a result of trawling, the Henwood and Stuntz study underestimates sea turtle mortality by a factor of as much as 3 (Figure 4-3).

A final underestimate results from the Henwood and Stuntz (1987) study having considered only shrimping effort in waters outside the coastal beaches. Because 33% of the total shrimping effort in 1987 and 1988 occurred in rivers, estuaries, and bays and because sea turtles (especially young Kemp's ridleys) are found in these waters, total mortality from the shrimp fishery could be higher than the Henwood and Stuntz estimates by a factor of as much as 1.6. That possibility is based on the assumption that the abundance of turtles is the same inside and outside and the assumption that a unit of effort is equal inside and outside; neither of those assumptions is precisely true (nor known, for that matter).

The limitations of the data and the criticisms of methods used do not detract from the basic findings of the Henwood and Stuntz study. With its assumption that all comatose turtles survive and its omission of all turtle capture and mortality estimates for inside waters, the approach taken by Henwood and Stuntz results in a marked underestimate of total sea turtle mortality associated with the shrimp fishery.

Relation Between Sea Turtle Stranding and Spatiotemporal Pattern of Shrimp Trawling in North Carolina The northern limit of the geographic zone of ocean shrimp trawling occurs at Ocracoke Inlet, North Carolina. Data compiled by Street (1987) on sea turtle stranding on ocean beaches in North Carolina exhibit a spatiotemporal pattern that closely matches that of trawl fishing in the ocean offshore of that state. South of Ocracoke Inlet, where offshore shrimp trawling continues from about May through September, 86% of the 545 sea turtle strandings observed in 1980-1986 occurred in those months. In contrast, north of Ocracoke Inlet, where no shrimp trawling occurs, but where a winter trawl fishery for flounder exists, 85% of the 456 sea turtle strandings recorded on ocean beaches in 1980-1986 occurred during the October-April period (Street, 1987). The spatiotemporal switch in the season and location of apparent sea turtle mortality suggests that shrimp trawling causes substantial mortality of sea turtles south of Ocracoke Inlet in North Carolina. The winter mortality of sea turtles to the north might be caused by groundfish trawling or by temperature shocks in the colder-water biogeographic province north of Cape Hatteras.

Other Fisheries, Discarded or Lost Gear, and Marine Debris

Turtles are caught and killed in finfish trawls, seines, pompano gill nets in Florida (pers. comm., L. Ehrhart, University of Central Florida, March 1990), various kinds of passive fishing gear (such as gill nets, weirs, traps, and long lines), lost fishing gear, and other debris. We conclude that the mortality associated with these and related factors is about one-tenth that associated with shrimp trawling (Table 6-2). Collectively, the nonshrimp fisheries constitute the second largest source of mortality of juvenile to adult sea turtles. That statement is based on the observations documented below by region.

The assessment of sea turtle mortality attributed to entanglement in stationary or fixed fishing gear is difficult, because of the disparity and discontinuity of reliable data. It is fair to assume that in some localities and with some types of fishing gear, entrapment and entanglement occur fairly often, but the resulting turtle deaths might not be as consistent. Most of the entangled or entrapped turtles are subadults and adults. Fishermen appear to be reasonably cooperative in efforts to set live sea turtles free. However, dead turtles are set adrift and might later be accounted for as strandings. The ratio of dead turtles set adrift to those counted as stranded is not adequately documented. If the approximate 4:1 ratio documented by Murphy and Hopkins-Murphy (1989) is considered valid, some total estimate of mortality can be made. On the basis of yearly stranding data with mortalities directly associated with encounters with fixed fishing gear, a yearly estimate of a maximum of 45-400 sea turtle deaths is reasonable. That is only a crude estimate; more research and monitoring are necessary to document and understand the interaction of sea turtles with fixed fishing gear.

Estimates of worldwide losses and discards of commercial fishing gear—including plastic nets, lines, and buoys—range from 1,350 to 135,000 metric tons of gear per year (Merrell, 1980; Welch, 1988). Recreational fishing in the United States is undoubtedly another important source of marine debris, including bait bags and lost and damaged gear (Pruter, 1987). NMFS recreational fishing statistics indicate that more than 81 million recreational fishing trips are made annually to marine waters (NMFS, 1986a,b).

It is especially difficult to document the deaths caused by this source or to separate them from deaths caused by fixed but unattended fishing gear. Yet, sea turtles (and other marine life) are particularly vulnerable to commercial fishing gear that has been lost or abandoned at sea. Such gear continues to catch and entangle marine life indiscriminately, causing injury, strangulation, starvation, and drowning (Carr, 1987; Laist, 1987; McGavern, 1989; Gregg, 1988). Deaths of green turtles, hawksbills, loggerheads, Kemp's ridleys, and leatherbacks have been caused by entrapment and entanglement in fishing gear (Mager, 1985). Monofilament line is the most common type of debris to entangle turtles. Other debris includes rope, trawl netting, gill netting, plastic sheets, and plastic bags. Fishing-related debris is involved in about 68% of all cases of sea turtle entanglement (O'Hara and Iudicello, 1987). Other stationary or passive fishing gear that has caused deaths of turtles includes pound nets, long lines, sturgeon nets, and nylon and monofilament gill nets (Van Meter, 1983). Leatherbacks and green turtles are prone to entangling their front flippers and heads in buoy ropes or discarded twine (O'Hara et al., 1986; O'Hara and Iudicello, 1987). The largest authenticated leatherback ever recorded became entangled in whelk-fishing lines and drowned; fishermen cut the dead turtle loose, and the carcass washed up the next day on a beach in Wales (Morgan, 1989).

Sea turtle entanglement in monofilament fishing line is a common problem. It is not usually related to active fishing; rarely is a fishhook reported attached to the line. In several cases, a turtle was entangled on line snagged on underwater structures or reefs, which caused constriction and necrosis of the limbs or drowning (O'Hara et al., 1986).

Balazs (1985) acquired reports of 60 cases worldwide of turtle entanglement involving monofilament line, rope, netting, and cloth debris. Of the 60 cases, 55 (92%) involved single animals, and 38% of all the turtles were dead or died later. Five species from 10 locations worldwide were reported; Kemp's ridleys were not included. Green turtles accounted for 19 (32%) of the 60 cases, and immature turtles were affected more often than adults in all the species represented except the leatherbacks. Only adult leatherbacks were reported entangled; immature leatherbacks are rarely reported anywhere. Monofilament line, with no fishhooks attached, accounted for 20 (33%) of the cases; segments or snarls of rope, 14 (23%); pieces of trawl or webbing, 12 (20%); and monofilament net, 8 (13%). Fishing-related debris was involved in 41 (68%) of all the cases.

New England

Leatherbacks and Kemp's ridleys become entangled in lobster gear (O'Hara et al., 1986). Balazs (1985) reported a dead leatherback from Rhode Island that had a longline hook embedded in its flipper, with rope attached. Although there have been no reports of sea turtle entanglement in gill nets in New England, reports of leatherbacks with cuts, severed limbs, or chafing marks suggest the possibility. Fretey (1982) published an extensive inventory of flipper injuries among leatherbacks in the large French Guiana nesting colony; some of these animals are known to come from feeding grounds in the northeastern United States. Balazs's (1985) compilation of worldwide incidence of sea turtle entanglement indicated that 11% of the 55 cases investigated involved monofilament net (O'Hara et al., 1986).

New York Bight

Turtle mortalities have resulted from lobster-pot lines and pound nets. Between 1979 and 1988, 58 stranded sea turtles reported in the New York Bight exhibited signs of entanglement with debris or inactive or fixed fishing gear. The Okeanos Ocean Research Foundation in New York reported two dead leatherbacks entangled in lobster gear in 1986 (O'Hara et al., 1986). Lobster-pot floatlines are a major source of entanglement, because they can be more than 180 m long in offshore waters and virtually undetectable below the surface. Six of 10 leatherbacks were caught in lobster-pot lines, and one entangled and drowned (Balazs, 1985; Sadove and Morreale, 1989).

In Long Island Sound, fixed pound-net gear traditionally captures the most sea turtles, predominantly Kemp's ridleys, but also green turtles, leatherbacks, and loggerheads (Morreale and Standora, 1989). The Okeanos Ocean Research Foundation has accumulated numerous reports of sea turtles, especially Kemp's ridleys, entrapped in pound nets in eastern Long Island. Surveyed fishermen indicate catching 10-20 turtles per year. That might be important, because more than 100 licensed fishermen were using pound nets in the region in 1986. It might not constitute a mortality problem, if the turtles are simply enclosed in the heart or head of the net until released, but deaths can occur if the turtles get tangled in the hedging or stringers (Balazs, 1985; Sadove and Morreale, 1989). Documented cases from 1986 include seven Kemp's ridleys, four loggerheads, and two green turtles captured; all but four were released alive (O'Hara et al., 1986). Balazs (1985) reported a leatherback that was found dead, tangled in rope. Debris in the water column or at the surface, such as floating line, can entangle turtles during normal activities, such as surfacing to breathe (Balazs, 1985; Sadove and Morreale, 1989).

Mid-Atlantic and Chesapeake Bay

The principal fishery-caused mortality in the mid-Atlantic and the Chesapeake Bay is in the pound-net fishery in the bay during the summer and the finfish trawl fishery for flounder off the coast in the winter. Documentation is best for the effects of the pound-net fishery. Other fishery-related mortality results from gill nets, crab-pot lines, and occasionally even rod-and-reel fishing. Some deaths in gill nets occur off Delaware (O'Hara et al., 1986).

Almost all turtle stranding during October and November in Virginia and adjacent waters of North Carolina occurred on the ocean front where heavy flounder trawling takes place off the coast. Some of the stranded turtles showed net marks and might have drowned in fish trawls. The seasonality of stranding in North Carolina north and south of Cape Hatteras implicates the flounder fishery as the source of mortality. Low-temperature deaths also might have contributed to the stranding. Further evaluation of this fall or winter mortality is warranted (Barnard et al., 1989).

An estimated 50-200 sea turtles strand from all causes in and around the Chesapeake Bay each year (Keinath et al., 1987; personal communication., D. Barnard and J. Keinath, Virginia Institute of Marine Science, October 1989). Stranding data for 1979-1988 were analyzed by Barnard et al. (1989) and D. Barnard and J. Keinath (pers comm., Virginia Institute of Marine Science, 1989). Of the turtles examined, 20% had definite net marks indicating death by pound net, gill net, or other fishing gear; 47% had no outward sign of injury or were very decomposed. The 20% figure is lower than previously estimated by Bellmund et al. (1987) and Keinath et al. (1987). Crab-pot lines and pound-net leads probably contributed to many of these deaths.

Stranding of dead turtles in and around Chesapeake Bay typically begins in mid-May. That pattern coincides with the deployment of pound nets in May. However, pound nets are in use through October, whereas strandings tend to cease by early July. The higher number of strandings early in the season might be related to the emaciated or weakened state of turtles entering the bay after a long migration (Bellmund et al., 1987; pers. comm., D. Barnard and J. Keinath, Virginia Institute of Marine Science, October 1989).

Many pound-net deaths might be related to the inability of sick or injured turtles to avoid fixed nets during periods of strong tidal flow (Musick, 1988; Barnard et al., 1989). Pound-net hedging or leaders with stringers produced the highest mortality rates for turtles, 0.7 per net, especially in strong currents. Pound-net leads composed of small mesh from top to bottom were associated with insignificant mortality rates. The turtle entanglement was 0.4 per net for open-water nets, compared with 0.1 per net for embayments and protected areas; the difference might be the result of the stronger currents in open water (Bellmund et al., 1987). In areas with weak currents, live turtles caught in pound nets apparently can move in and around the nets without becoming entangled (Bellmund et al., 1987), as evidenced by live turtles marked and released from one net that have later been recaptured in the same or a nearby net and by the observation of a few loggerheads crawling out over the head netting as the net was being worked (Lutcavage, 1981).

It is unlikely that stranded turtles without visible constrictions were killed in pound nets. Entangled turtles in pound nets die and begin to decompose in situ; they do not drift free to strand on shore (Bellmund et al., 1987). None of five dead turtles entangled in pound-net hedging during 1984 came loose over 5 weeks. However, the rotting turtle eventually bloats; as it decomposes, it tears free, floats away, and strands (Lutcavage, 1981). One dead and marked loggerhead from pound-net hedging stranded 5 days later 10 km from the net.

Various reports have assessed the sources of mortality of dead sea turtles stranded on inshore beaches and shores in and around the Chesapeake Bay. A total of 645 dead turtles, including 527 loggerheads and 28 Kemp's ridleys, stranded between May 1979 and November 1981. Necropsies of some loggerheads implicated enteritis and drowning (Lutcavage and Musick, 1985). A sample of 71 turtles from 1979 included 25 with a determinable cause of death; seven of the deaths were caused by pound nets. Of the 57 turtles sampled in 1980, 21 had a determinable cause of death, and 19 deaths were caused by pound nets. In addition to pound-net deaths, one turtle died in a haul seine, one after being caught on a long line, and two in crab-pot lines (Lutcavage, 1981). Of the 124 turtles sampled in 1981, 11 had determinable deaths, and four deaths were caused by pound nets (Lutcavage and Musick, 1985). Confirmed netting deaths from 1979 to 1983 numbered 53 (19% of the determinable deaths); only four turtles (1.4%) died as a result of non-net fishing gear. Of the 83 dead stranded turtles examined in 1984, 10 (12%) had evidence of constriction, and 20 (24%) were in pound or gill nets (Bellmund et al., 1987). Definite net-related deaths in the Chesapeake Bay during some summers from 1979 to 1984 ranged from 3% to 33% of the total number of stranded turtles (Lutcavage and Musick, 1985).

Early reports indicated that the cause of death could be determined for about half the 980 stranded sea turtles recorded between 1979 and 1987 and that almost 40% could be attributed to entanglement in gill or pound nets (Keinath et al., 1987). However, reanalysis of the data available for 1979-1988 determined that fewer turtle deaths (approximately 20%) could be definitely attributed to entanglement in pound or gill nets, or other fishing gear (Barnard et al., 1989).

South Atlantic

Sea turtle deaths other than those caused by shrimp fishing have occurred in the south Atlantic in association with oceanic gill nets, large ocean set nets, and tuna and billfish long lines.

Turtle mortality associated with gill-net fisheries in the Carolinas starts in early spring and is maximal in April. The South Carolina Wildlife and Marine Resources Department reported that oceanic gill net fisheries for Atlantic sturgeon, shad, and shark have caused the deaths of loggerheads, Kemp's ridleys, and green turtles (pers. comm., S. Murphy, S.C. Wildlife and Marine Resources, 1989). In 1980-1982, about 217 turtles stranded in the early spring in connection with large ocean nets used to catch Atlantic sturgeon. In 1983-1985, the sturgeon season was closed in mid-April, and the carcass count decreased to about 106 turtles. In 1986, there was no sturgeon season, and only about 18 turtles died in the spring. Illegal drift nets for the shad fishery and shark fishery were probably responsible for most of the 36 carcasses reported in May 1989, including eight leatherbacks.

Sea turtles are caught infrequently on long lines in the gulf and Atlantic (Swordfish Management Plan, 1985). On the basis of data from the 1979 Japanese long-line observer program, 12 turtles (including two leatherbacks) were caught in the gulf and 17 (including nine loggerheads) were caught in the Atlantic. During 1980, the same observer program reported 10 turtles captured. The greatest number were captured in January-March in the gulf and in September-January in the Atlantic. Seven percent of the turtles captured died in gulf long-line fishery and 30% in incidental captures in the Atlantic (O'Hara et al., 1986). One unidentified turtle in 1987 and one leatherback in 1988 were hooked, as reported by observers on Japanese long-line vessels fishing in the northwest Atlantic fishery conservation zone (FFOP, 1988, 1989). Leatherbacks tend to get hooked (either in the mouth or the flipper area), whereas loggerheads are prone to entanglement in the ganglion lines attached to the main line.

In Florida, there were five recent confirmed sightings by divers of sea turtles entangled in monofilament fishing line on reefs and wrecks (pers. comm., J. Halusky, N.E. Florida Sea Grant Extension, May 1989). Two of them were rescued and released, and three were dead.

Balazs (1985) reported 10 cases of turtle entanglement in Florida in 1978-1984: one green turtle, alive; five loggerheads, including three dead; and five hawksbills, alive. Balazs also reported a dead loggerhead in Georgia. Of the 11 cases, six involved monofilament fishing line, two involved rope, two involved gill or other netting, and one involved both line and netting.

Gulf Coast

A study along the Texas coast during 1986 and 1987 encountered entanglement of 25 turtles in discarded net and monofilament line. Entanglement was identified as the probable cause of death of seven; the remainder were stranded alive. Nine of the 25 turtles were Kemp's ridleys, and the others were loggerheads, hawksbills, green turtles, and leatherbacks. The turtles were entangled in fishing line and hooks, shrimp trawls, onion sack, net and rope, tar, crab trap, and trot line. The study concluded that the probability that a sea turtle in Texas coastal waters would come into contact with marine debris is high, and that commercial and recreational fishermen and their discarded gear were responsible for most of entanglements (Plotkin and Amos, 1988; Ross et al., 1989).

Balazs (1985) reported five entangled turtles in Texas in 1977-1983, including one live green turtle, three live hawksbills, and one dead hawksbill. Four of the entanglements involved monofilament fishing line, the other a piece of plastic onion bag. Amos (1989) reported that, in 77 recorded strandings of hawksbills in Texas since 1972, the incidence of entanglement in plastic was high—22% of those in which such information was recorded. The most common form of entanglement occurred when turtles' necks or limbs were caught in woven plastic produce sacks. Monofilament fishing line wrapped around limbs has also been recorded. No entanglements of recent posthatchlings have been noted, only entanglements of yearlings.

An anecdote from Paul Raymond of the NMFS Law Enforcement Division provides a dramatic example of the problem. An abandoned pompano trammel net (three panels) of monofilament was seized on October 16, 1989, off the beach (near shore) near Wabasso, Florida (Indian River County). It had been set 6 days before and left unchecked. In it were 10 juvenile green turtles and one juvenile loggerhead, all entangled and drowned. Pompano trammel nets are tethered in very shallow water near shore by fishermen in small boats. The industry is not well organized or documented as to size, season, or distribution. Nets set inshore (behind the coastal regulation lines) must be attended, but that is not required by Florida for nets outside the line. The net in question here had been abandoned. An unattended but not abandoned net can also kill turtles.

Dredging

Dredging of harbors and entrance channels can kill sea turtles (Hopkins and Richardson, 1984). A comprehensive survey of records and project reports recognized 149 confirmed incidents of sea turtles entrained by hopper dredges working in two shipping channels from 180 to 1990 (Table 6-6) (pers. comm., J.I. Richardson, University of Georgia, April 1990). Only verifiable records of fresh kills or live turtles were included in this table, explaining the slight difference in total counts between this survey and other reports (Rudloe, 1981; Joyce, 1982). Three species of sea

TABLE 6-6 Reported sea turtle incidents by species during dredging activities from 1980 to 1990.

Site	Year	Loggerhead	Green Turtle	Unidentified*	Total
Cape Canaveral	1980	50	3	18	71
Entrance Channel	1981	1	1	1	3
	1984-85	3	0	6	9
	1986	5	0	0	5
	1988	8	2	18	28
	1989-90	0	6	1	7
Totals		67	12	44	123
King's Bay	1987-88†	7	1	1	9
Entrance Channel,	1988	3	0	2	7 ‡
Georgia and Florida	1989	9	0	1	10
Totals		19	1	4	26

*Fragments of sea turtle carcasses not identified to species. It is assumed that most are loggerheads.
†Initial construction dredging for Trident submarine base.
‡Two Kemp's ridleys caught in 1988 at Kings's Bay, Georgia.

Source: Richardson, 1990.

turtles were taken, including two individuals of the endangered Kemp's ridley. Although some entrained specimens were identified, it is estimated that 90% or more of the incidents involved the loggerhead. Nearly all sea turtles entrained by hopper dredges are dead or dying when found, but an occasional small green turtle has been known to survive.

Entrapment and death of turtles by hopper dredges first became an issue of concern at the Port Canaveral Entrance Channel, Florida, in 1980 after unusually high concentrations of loggerheads were noted in the area (Carr et al., 1981). Seventy-seven loggerheads were reported killed in 1980 during the removal of 2.5 million cubic yards (1.9×106 m^3) of sediment from the channel (Rudloe, 1981; Joyce, 1982). The rate of turtle take varied among dredges, ranging from 0.038 turtle entrained per hour (dredge *McFarland*) to 0.121 turtle entrained per hour (dredge *Long Island*) (Joyce, 1982). The very high number of turtles taken was not repeated in subsequent years for several reasons. First, the *Long Island,* because it seemed to pose the greatest threat, was transferred immediately to other areas. Second, a program of gear modification to the drag heads was initiated at that time. Finally, the loggerheads did not seem to use the Canaveral Channel in the same numbers in later years. By 1989, the rate of sea turtle capture in surveys in the channel were about one-tenth the rates recorded in 1978-1983 (pers. comm., A. Bolten, University of Florida, 1989).

Although the most serious loss of turtles in hopper dredges occurs within the Port Canaveral Entrance Channel, smaller numbers have been taken at King's Bay Entrance Channel. Twelve turtles (10 juvenile loggerheads, one adult loggerhead, one juvenile green turtle) were taken during some 20,000 hours of construction dredging (Slay and Richardson, 1988). The rate of capture was less than 0.001 turtle per dredge hour.

The loss of turtles to hopper dredges in other entrance channels is not yet known, but other entrance channels from North Carolina to Florida will be surveyed by NMFS to assess any potential effects on sea turtles (pers. comm., J. Richardson, University of Georgia, 1990). Data are being gathered through additional observer programs to answer the question, and the numbers of sea turtles taken are expected to be considerably smaller than observed at Port Canaveral. The data are not available, but it would not be unusual for 1,000 hours of maintenance dredging to be needed per channel per year.

Collisions with Boats

Another source of mortality to sea turtles associated with human activity is collision with vessels. The regions of greatest concern are those with high concentrations of recreational-boat traffic, such as the southeastern Florida coast, the Florida Keys, and the many shallow coastal bays in the Gulf of Mexico. Of the turtles stranded on the gulf and Atlantic coasts of the United States, 6% of 1,847 strandings in 1986, 7% of 2,373 in 1987, and 9% of 1,991 in 1988 had boat-related injuries for an average of about 150 turtles per year (Schroeder, 1987; Schroeder and Warner, 1988; Teas and Martinez, 1989). In most cases, it was not possible to determine whether the injuries resulted in death or were postmortem injuries.

In the Chesapeake Bay region, boat-propeller wounds accounted for approximately 7% of the deaths of sea turtles stranded in 1979-1988 whose causes were determinable (Barnard et al., 1989), or about five to seven turtles per year.

If we assume that half the boat-collision injuries documented by the STSSN were the primary causes of death of the stranded sea turtles in 1986-1988, and only about 20% of the dead turtles wash ashore, about 400 turtles are killed by boat collisions each year along the gulf and Atlantic coasts of the United States outside of coastal beaches. That estimate might be low, because the strandings include only the ocean beaches (boat collisions with turtles also occur in inside waters), and an animal with an open wound has an increased probability of predation and thus a further reduction in probability of stranding. The estimate might be high, because more than half of the turtles might have been hit when they were already dead from other causes and were floating.

Petroleum-Platform Removal

The use of explosives in removal of petroleum structures became controversial with respect to turtle mortality in 1986. From March 19 to April 19, 1986, 51 turtles, primarily Kemp's ridleys, were found dead on beaches of the upper Texas coast. Ten petroleum structures in the nearshore area of the strandings had been removed with explosives during the period. Shrimping was at a seasonal low, and circumstantial evidence suggested that at least some of the strandings were due to underwater explosions used in removal of the structures (Klima et al., 1988). Further evidence of the serious effects of the explosions included the stranding of 41 bottlenose dolphins (*Tursiops truncatus*) and large numbers of dead fish (Klima et al., 1988).

After those incidents, attention focused on the possible effects of petroleum-platform removal. In July 1986, 11 sightings of at least three turtles (two loggerheads and one green turtle) occurred during the removal of a platform 30 miles south of Sabine Pass, Texas. What appeared to be a dead or injured turtle drifting with the current 10 feet below the surface was reported 1.5 hours after detonation of explosives (Gitschlag, 1989). Six sightings of loggerheads were reported at five other removal sites, and a green turtle was observed at another location. Those sightings and strandings resulted in a consultation under Section 7 of the Endangered Species Act of 1973 between NMFS and the Minerals Management Service (MMS). Oil and gas companies wishing to use underwater explosives were thereafter required to submit permit requests to MMS. Obtaining a permit requires use of qualified observers to monitor sea turtles near platforms and in some cases to remove turtles to a safe location away from the potential impact of explosive charges.

Data collected by NMFS since 1986 support an association between turtles and some offshore platforms. Divers have reported that turtles commonly associate with offshore structures (Rosman et al., 1987). Gitschlag (1989) reported that 36 turtle sightings near platforms scheduled to be removed were made during 1987-1988 by the NMFS observer program. Another 30 turtles were observed during that period at structures not scheduled for removal (personal communication, G. Gitschlag, NMFS, 1989). A recent NMFS observer effort indicated that turtle concentrations near a petroleum platform could be large. Twelve turtles were collected and removed from one structure off Texas in September 1989 (pers. comm., G. Gitschlag, NMFS, 1989).

Additional reports confirm the association of turtles with offshore platforms. Lohoefener (1988) used aerial surveys and found hard-shelled sea turtles (cheloniids) to be associated with platforms offshore of the Chandeleur Islands (Louisiana), although their study did not indicate an association of sea turtles with platforms in the western Gulf of Mexico. They determined the daytime probability of one or more cheloniids near a platform off the Chandeleur Islands to be about 0.27 within 500 m of the structure, 0.50 within 1,000 m, and 0.65 within 1,500 m. West of the Mississippi River, the probability of one or more cheloniids within 500 m of a randomly selected platform would be about 0.04, within 1,000 m about 0.08, and within 1,500 m about 0.13. Only larger turtles and only turtles on or near the surface are usually seen by aerial surveys, so the figures given should be considered low.

Although information on association of sea turtles with energy platforms is sparse, the potential for mortality must be considered genuine. It is difficult to document a cause-effect relation between turtle deaths and offshore explosions, because no dead animals have been recovered at removal sites and freshly killed turtles sink and might drift a long way by the time putrefaction causes them to float. Association of turtles with the structures is not random; platforms apparently provide a resting place or a location where food is readily available (Klima et al., 1988). From March 1987 through 1988, 69 platforms and 39 caissons or other single-pile structures were removed in gulf waters of Louisiana and Texas. MMS estimated that there were 3,434 platforms in the federal outer continental shelf as of December 1986 and predicted that 60-120 structures would be removed each year for the next 5 years (MMS, 1988). Continuing research should identify more specifically the negative effects of explosive removal of offshore structures. Safeguards for protection of turtles near structures scheduled for removal are essential.

To estimate the numbers of sea turtles that might be killed or injured by explosions in the future, we assumed that the injury and mortality zone will extend no farther than 1,000 m from the structure being removed (Klima et al., 1988). For the Chandeleur Islands area, where the highest densities were seen, Lohoefener et al. (1988) used aerial surveys and estimated a 0.5 probability that a turtle would be visible within 1,000 m of a given structure during the day. If about 100 platforms in gulf waters of Louisiana and Texas will be removed each year over the next 10 years, a total of 8-50 turtles each year could be killed or injured without protective intervention. That estimate is biased downward for two reasons: first, an aerial survey samples only during the day, when turtles are known to forge away from resting sites. Second, turbidity in the Gulf of Mexico may reduce visibility from the air, especially west of the Mississippi. Yet, Klima et al. (1988) estimated higher densities of turtles in this region during the observer programs. If only half of the turtles are seen during aerial surveys, the estimate could reach as high as 100 turtles possibly affected each year over the 10-year period.

Other uses of explosives also might have an effect. Petroleum seismographic exploration and military maneuvers can use explosives. Their impact on turtle mortality has not been measured, but it might exist. In contrast, turtles nesting in areas adjacent to military bombing activities (e.g., on eastern Vieques Island, Puerto Rico) might actually benefit, because the control of human access and the danger of unexploded rounds greatly reduce the presence of egg poachers (pers. comm., P. Pritchard, Florida Audubon Society, October 1989).

Entrainment of Sea Turtles in Power-Plant Intake Pipes

Sea turtles can become entrained in intake pipes for cooling water at coastal power plants. The best-documented case is that of St. Lucie unit 2 in southeastern Florida. At that facility, nets are constantly set and monitored in the intake canal to remove sea turtles. In 1976-1988, 122 (7.5%) of the 1,631 loggerheads and 18 (6.7%) of the 269 green turtles entrapped in the canal were found dead, for an average of about 11 turtles per year (Applied Biology Inc., 1989a). Four Kemp's ridleys were found dead during the same period. No dead leatherback or hawksbill has been found there (Applied Biology Inc., 1989a). Deaths resulted from injuries sustained in transit through the intake pipe, from drowning in the capture nets, and perhaps from causes before entrainment.

At four other power plants in Florida (Port Everglades, Turkey Point, Cape Canaveral, and Riviera Beach), 21 turtles (loggerheads, green turtles, and one hawksbill) were entrained in the systems from May 1980 through December 1988. Of the 21, seven were found dead (four of which were loggerheads), for an average of about one per year. At the Port Everglades plant, 25-30 hatchlings were also entrained in the system, and a few of them died (Applied Biology Inc., 1989b).

Other turtle deaths at coastal power plants have been reported in New Jersey (Eggers, 1989), North Carolina, and Texas (pers. comm., T. Henwood, NMFS, 1989; pers. comm., B. Schroeder, Florida Department of Natural Resources, 1989). They were sporadic and apparently involved few turtles. For example, the Delaware Bay Power Plant in New Jersey entrapped 38 turtles in 9 years—26 loggerheads (18 dead) and 12 Kemp's ridleys (six dead), for an average of about three per year.

Two factors cause an unusually high entrainment rate at the St. Lucie unit 2 power plant in Florida. First, the continental shelf is narrow in that area, and that seems to cause the normally high density of turtles passing along the coast to be concentrated near the shore, where the coolant-water intake tube is. Second, that part of the coast appears to be on the main coastal migratory route for turtles in the region. Therefore, mortality rates for this power plant should be considered separately. A total mortality estimate of about 11 turtles per year might be expected in the future at current population densities: about 9.4 loggerheads, one green turtle, and 0.3 Kemp's ridley.

For other power plants, far less is known. According to the Edison Electric Institute (1987), 98 power-generating facilities use ocean or estuarine water for their cooling systems along the gulf and Atlantic coasts of the United States. If we assume that rates of turtle capture from the five power plants discussed above (excluding the St. Lucie facility) are typical for the remaining coastal facilities between New York and Texas, we can estimate an annual mortality of 48 loggerheads and 13 Kemp's ridleys (loggerheads, 98 power plants × 0.48 per year; Kemp's ridleys, 98 power plants × 0.13 per year). Adding in the estimates from the St. Lucie plant raises the loggerhead to 57 per year and Kemp's ridley to 13 per year. An important consideration for the future is that, as turtle populations increase, we would expect an increase in the number of animals entrained in the facilities, and as human populations increase, more power plants might be built.

Directed Take

Directed take of sea turtles and their eggs is illegal in the United States and along the Caribbean and gulf coasts of Mexico. Some illegal take does occur in the United States and Mexico, but the numbers are probably negligible. Loss of eggs and adult Kemp's ridleys at Rancho Nuevo is minimal, because protection has been provided. Although directed take of sea turtles is widely considered to affect populations, at least locally (Pritchard, 1980), the committee was unable to quantify the extent of the problem.

Toxicology

Tissues and eggs from several species of sea turtles in the southeastern United States, Ascension Island in the South Atlantic, the coast of France, and other geographic regions have been analyzed for organochlorine compounds, heavy metals, hydrocarbons, and radionuclides (Hillestad et al., 1974; Thompson et al., 1974; Stoneburner et al., 1980; Clark and Krynitsky, 1980, 1985; Witkowski and Frazier, 1982; Bellmund et al.,

1985). Turtles were found to be contaminated to various degrees in all the studies cited. However, because of the lack of data on physiological effects of the pollutants in sea turtles, their effect on survival cannot be estimated. Additional studies are needed to determine extents of contamination and the physiological effects of the contaminants.

Ingestion of Plastics and Other Debris

About 24,000 metric tons of plastic packaging is dumped into the ocean each year (Welch, 1988). Nationwide 10-20% of beach debris is expanded polystyrene foam and 40-60% is other plastic (McGavern, 1989). An estimated 1-2 million birds and more than 100,000 marine mammals and sea turtles die from eating or becoming entangled in plastic debris each year, including netting, plastic fishing line, packing bands, and styrofoam (Welch, 1988; McGavern, 1989; Sanders, 1989).

Sea turtles ingest a wide variety of synthetic drift items, including plastic bags, plastic sheeting, plastic particles, balloons, styrofoam beads, and monofilament fishing line. Specific reports have been related to green turtles in Hawaii, Florida, and Texas; loggerheads in Georgia, Florida, Texas, and Virginia; hawksbills in Florida and Hawaii; and leatherbacks in New York, New Jersey, Massachusetts, and Texas (Wallace, 1985; O'Hara et. al., 1986). Turtles can mistake plastic bags and sheets for jellyfish or other prey. Ingestion of those items can cause intestinal blockage; release toxic chemicals; reduce nutrient absorption; reduce hunger sensation; inhibit feeding and mating activity; diminish reproductive performance by leaving the turtle unable to maintain its energy requirements and cause suffocation, ulceration, intestinal injury, physical deterioration, malnutrition, and starvation (Wehle and Coleman, 1983; Wallace, 1985; O'Hara et al., 1986; Bryant, 1987; Farrell, 1988; Gramentz, 1988; Welch, 1988; McGavern, 1989).

Absorption of toxic plasticizers (such as polychlorinated biphenyls) is also possible as a result of ingestion. Some plasticizers can concentrate in tissues, and the toxic ingredients can cause eggshell thinning, tissue damage, and aberrant behavior (Wehle and Coleman, 1983; O'Hara et al., 1986).

Plastic bags blocked the stomach openings of 11 of 15 leatherbacks that washed ashore on Long Island during a 2-week period. Ten had four to eight quart-sized bags, and one had 15 quart-sized bags (*San Francisco Chronicle*, 1983; Balazs, 1985; O'Hara et al., 1986). In South Africa, Hughes extracted a ball of plastic from the intestine of an emaciated leatherback; when unraveled, it measured 9 × 12 ft, or 2.7 × 3.7 m (Balazs, 1985; Coleman, 1987). In September 1988, the largest leatherback ever recorded (914 kg) was found dead on a beach in Wales. The cause of death was listed as drowning due to entanglement, but a tightly compacted piece of plastic (15 × 25 cm) blocked the entrance to the small intestine and might have contributed to death (Eckert and Eckert, 1988).

Accumulation of pollutants and plastic debris found in sargassum driftlines might be a source of mortality of turtles through ingestion (Mager, 1985). Floating debris is concentrated by natural processes along lines of convergence between discrete water masses, in the core of major current gyres, or on beaches and submerged rocky outcrops. Driftlines along margins of small temporary eddies or areas of downwelling can accumulate floating debris and provide feeding areas for turtles. Young turtles are passive migrants in offshore driftlines and can contact buoyant debris.

In 1979-1988 in the New York Bight area, necropsies were performed on 116 sea turtles. Various amounts of synthetic materials were found in 10 of 33 leatherbacks, three of 35 loggerheads, one of four green turtles, and none of 44 Kemp's ridleys. Most prevalent were plastic bags, small pieces of plastic sheeting, monofilament line, small pieces of variously colored plastic, and numerous small polystyrene balls. There was strong evidence in some animals that ingestion of synthetic materials caused their deaths. There is little information on the residence times and cumulative effects of synthetic materials in marine animals. These observations are not well suited to quantify the frequencies of ingestion (Sadove and Morreale, 1989).

Studies conducted along the Texas coast in 1986-1988 documented the effects of marine debris on sea turtles (Plotkin and Amos, 1988; Stanley et al., 1988; Plotkin, 1989). They were significantly affected by ingestion of, and to a smaller extent entanglement in, marine debris. Necropsies of Kemp's ridleys, loggerheads, and green sea turtles revealed that the intestines of at least 65 of 237 turtles examined contained marine debris, such as plastic bags, styrofoam, monofilament fishing line, polyethylene beads, aluminum foil, tar, glass, and rubber. In a 22-month study, plastic was found in nearly 80% of the turtle stomachs that contained debris and in turtles from about 97% of the beaches surveyed (Stanley et al., 1988). All five species found in the Gulf of Mexico had eaten or were ensnared by debris.

Reports of debris ingestion by species indicated that green turtles had the highest incidence (32%) followed by loggerheads (26%), leatherbacks (24%), hawksbills (14%), and Kemp's ridleys (4%). For all species except the leatherback, immature turtles were involved more frequently than adults. The distribution of debris types was as follows: plastic bags and sheets (32.1%), tar balls (20.8%), and plastic particles (18.9%).

NMFS scientists, on the basis of the results of autopsies conducted since 1978, estimated that one-third to one-half of all turtles have ingested plastic products or byproducts (Cottingham, 1988).

Mortality associated with ingestion of plastics and debris cannot be accurately quantified from available data. Of the turtles examined, green turtles ingested plastic debris most frequently, followed by loggerheads and leatherbacks. Research is needed to develop accurate postmortem techniques to determine the role of plastic ingestion on turtle deaths. However, many reported stranded turtles are in an advanced state of decomposition, so it is difficult to determine exact causes of death, although indigestible stomach contents might still be identifiable. It is possible that the enactment of Annex V of the International Convention for the Prevention of Pollution from Ships (called MARPOL for "marine pollution") might affect the amount of plastic that sea turtles are likely to encounter in the future; but, considering the life span of plastic and the amount already present in the oceans, the possible deleterious effects of plastic on sea turtles and other wildlife will be present for generations to come.

SUMMARY

Sea turtles are susceptible to human-caused deaths through their entire life, from nesting females, eggs, and hatchlings on beaches to juveniles and adults of both sexes in offshore and inshore waters.

The committee found that the most important source of mortality on eggs and hatchlings at present on U.S. beaches is from non-human predators, whose abundance is often associated with human disturbance, but other factors are beach development, directed take, beach vehicles, and beach lighting. The most important source of mortality for juveniles to adults in the coastal zone is shrimp trawling. Other factors judged to be of significance for juveniles and adults are other fisheries and entanglement in lost fishing gear and marine debris.

Order-of-magnitude estimates of human-caused mortality on juvenile to adult loggerheads and Kemp's ridleys were made by the committee. Shrimp trawling accounts for 5,000-50,000 loggerhead and 500-5,000 Kemp's ridley mortalities per year. Other fisheries and discarded fishing gear and debris account for 500-5,000 loggerhead and 50-500 Kemp's ridley mortalities. Dredging, collisions with boats, and oil-rig removal each account for 50-500 loggerhead and 5-50 Kemp's ridley deaths. Entrainment in electric power plants and directed take each account for fewer than 50 turtle deaths per year. Based on the committee's evaluation, about 86% of the human caused mortalities on juveniles and adults result from shrimp trawling. The committee recognized the possible effects of plastic ingestion and marine debris but was unable to quantify them.

The strong evidence that shrimp trawling is the primary agent for sea turtle mortality caused by humans comes from five lines of analysis and information. First, the proportion of sea turtles caught in shrimp trawls that are dead or comatose increases with an increase in tow time from 0% during the first 50 minutes to about 70% after 90 minutes. Second, the numbers of turtles stranding on the coastal beaches consistently increased in a steplike fashion when the shrimp fishery opened in South Carolina and Texas and decreased when the fishery closed in Texas. Because the openings and closings were on different dates in different years, the change in strandings can be ascribed to the fishery rather than to date per se. The change in stranding rate indicates that 70 to 90% of the turtles stranded at those times and places were killed in shrimp trawls. Based on analysis of data from loggerheads, these stranded turtles were also in the life stages with the highest reproductive values. Third, loggerhead nesting populations are declining in Georgia and South Carolina where shrimp fishing is intense, but appear to be increasing farther south in central to southern Florida where shrimp fishing is low or absent. Fourth, the estimate in the literature of 11,000 loggerheads and Kemp's ridleys killed annually by shrimp trawling was judged by the committee to be an underestimate, possibly by a factor of three to four, because that estimate accounted for neither mortality in bays, rivers, and estuaries nor the likely deaths of most comatose turtles brought onto the deck of shrimp trawlers. Many of the comatose turtles will die even when released back into the water. Fifth, in North Carolina, turtle stranding rates increase in summer south of Cape Hatteras when the shrimp fleet is active and north of Cape Hatteras in winter when the flounder trawling is active.

7

Conservation Measures

The endangered or threatened status of sea turtle species in U.S. waters dictates aggressive and comprehensive management plans to expedite population recoveries. The immediate goal of any management scheme must be to arrest population declines. The ultimate goal is to establish conditions that permit breeding populations to increase numbers to some level at which a species is no longer at appreciable risk of extinction. Most strategies for achieving those goals are in broad, nonexclusive categories: strategies to increase the supply of animals and strategies to reduce causes of death so that animals in the system have a better chance of entering and remaining in the breeding population.

Natural mortality factors, except those affecting eggs and hatchlings on beaches, typically are difficult, if not impossible, to manipulate; mortality factors that result from human activities are more amenable to management. Strategies to increase reproduction and reduce mortality will be discussed in this section after we describe the general rationale and objectives of recovery plans.

RATIONALE AND OBJECTIVE OF THE RECOVERY PLAN

The Endangered Species Act of 1973 (Public Law 93-205) provides for the conservation, protection, and propagation of species of wild fauna and flora actually or potentially in danger of becoming extinct. All sea turtles in U.S. waters have been listed as either endangered or threatened.

An endangered species is "any species, subspecies, or distinct population of fish, or wildlife, or plant which is in danger of extinction throughout all or a significant portion of its range." A threatened species is "any species, subspecies, or distinct population of fish or wildlife, or plant which is likely to become an endangered species within the foreseeable future throughout all or a significant portion of its range."

The status of discrete breeding populations of listed species must be reviewed every 5 years, and recommendations, if warranted by the biological data, for delisting or reclassification must be made to the secretaries of the Departments of the Interior and Commerce, who jointly administer jurisdictional responsibilities for sea turtles.

The leatherback and hawksbill were listed as endangered throughout their ranges on June 2, 1970. The Kemp's ridley was listed as endangered on December 2, 1970. The green turtle was listed on July 28, 1978, as threatened, except for the breeding populations of Florida and the Pacific coast of Mexico, which were listed as endangered. The loggerhead was listed on July 28, 1978, as threatened wherever it occurs. Those sea turtles were listed because, to various degrees, their populations had declined as a result of human activities. Many of their nesting beaches had been affected by encroachment of the human population into coastal habitats. Sea turtle populations had been reduced by uncontrolled harvesting for commercial purposes and by deaths incidental to such activities as commercial fishing. In many cases, regulations did not increase conservation efforts.

The ESA requires the preparation of a recovery plan for each listed species, unless the department secretaries find that a recovery plan will not further the recovery of a particular species. It allows for the formation of recovery teams responsible for developing recovery plans. The objective of a plan is the survival and eventual recovery of a listed species or population, so that it can be removed from the endangered or threatened list.

The recovery plan for sea turtles prepared by the Marine Turtle Recovery Team was approved by NMFS in 1984. The plan, based on the best available information, recognized the difficulty of managing species that migrate outside U.S. jurisdiction and are commercially exploited in other countries. Within U.S. jurisdiction, the plan recommended management practices to enhance production on nesting beaches and to reduce mortality at sea and on land. Updated recovery plans for each species will be released in 1990. In general, recommendations of this committee are consistent with those of the 1984 recovery plan for marine turtles, but more recent or continuing declines of some species caused the committee to enlarge its set of recommendations.

DESCRIPTION OF CONSERVATION MEASURES

In this and the following sections, the committee relied on draft recovery plans for material on beach management and education.

Increasing or Maintaining the Supply of Sea Turtle Eggs and Hatchlings

Some approaches are aimed at increasing the supply of sea turtle hatchlings that enter the ocean system and eventually join the breeding population. Management measures range from protection of habitats (particularly critical nesting habitats) to captive breeding and programs of delayed release of young turtles.

Increasing Protection of Critical Nesting Habitats

Land Use

A joint state, federal, and private effort is under way to provide permanent protection for 15 km of the approximately 34 km of high-density sea turtle nesting habitat between Melbourne Beach and Wabasso Beach, on the Atlantic coast of east-central Florida (Possardt and Jackson, 1989). In that area, loggerhead nesting averages 475 nests/km in Brevard County and 140 nests/km in Indian River County. About 35-40% of green turtle nesting and 25% of loggerhead nesting in the southeastern United States occurs in southern Brevard County and northern Indian River County. Purchase of undeveloped beach property along that stretch (acquisition of "in-fee title") is the best way to conserve it. Obtaining conservation easements on undeveloped beach property will not be sufficient to ensure long-term protection. Only if the "in-fee titles" are acquired will the continued protection of this critical nesting beach be ensured, because only then will full control rest with land management authorities. If condominiums and other structures are built behind the nesting beach, the beach will eventually be lost as a result of storms or rising sea levels, because it will not be able to migrate naturally.

Erosion

Efforts to mitigate the effects of erosion usually consist of transferring nests to higher sites on a dune or into a hatchery. Those have now become common practices on many nesting beaches throughout the nesting range of the loggerhead and other species. Relocation projects are authorized under state and federal permits.

Beach Armoring

The destructive use of sea walls and other means of beach armoring (see Chapter 6) continues as a rising sea level erodes private property and threatens existing homes and other human structures. States have not taken the drastic action of removing or prohibiting construction of sea walls, but efforts are under way to adjust zoning so as to avoid the need for beach armoring on currently undeveloped lands. Most zoning ordinances are at the county level.

Beach Nourishment

Beach nourishment is less destructive of sea turtle nesting habitat than is beach armoring, but it can cause problems for nesting females and nests if not done properly. NMFS regulates beach-nourishment projects in behalf of the sea turtles and requires mitigation measures through the mechanism of Section 7 ("Consultation") of the ESA. Such regulation is possible, because nearly all beach-nourishment projects receive federal aid and therefore require endangered-species consultations. In Florida, much beach nourishment occurs in the summer, and nests must be moved from the beach before nourishment (e.g., Wolf, 1989). The quality of nourishment material must be acceptable to nesting sea turtles (Nelson and Dickerson, 1989a). The policy of beach nourishment is under continuing review by the Fish and Wildlife Service, Army Corps of Engineers, and the Florida Department of Natural Resources.

Increasing Protection of Nesting Adults, Eggs, and Hatchlings

Artificial Lighting

Considerable progress has been made in developing artificial lighting that does not compromise the efforts of nesting turtles or the emergence of the hatchlings (Dickerson and Nelson, 1988, 1989; Nelson and Dickerson, 1989b), particularly low-pressure sodium lights that appear to have a minimal effect on sea turtle orientation. Low-pressure sodium lights still prompt some concerns, and research continues.

In the absence of acceptable lighting, many states, counties, and towns are making progress in mitigating the effects of light (pers. comm., L.

Shoup and R. Wolf, Alachua County Department of Environmental Services, Gainesville, Florida, 1987; pers. comm., R. Ernest, Applied Biology, Inc., Jensen Beach, Florida, 1990). In Florida, lighting ordinances have been passed in several counties and are being considered in others (pers. comm., J. Huff, Florida Department of Natural Resources, 1989).

Beach Cleaning, Pedestrian and Vehicular Traffic, and Recreational Equipment

Beach-cleaning equipment, pedestrian traffic, off-road vehicles, and other human activities disturb nesting sea turtles considerably and can destroy eggs and hatchlings. Off-road vehicles are regulated on many beaches, but are still allowed on the beaches of North Carolina, Georgia, parts of northern Florida, and Texas. The pedestrian traffic problem is often solved by moving nests out of the way of beach access ramps or marking their presence in such a way that beach users will avoid them. On the beach at Boca Raton, small screen cages are placed over the nests (pers. comm., R. Wolf, Alachua County Department of Environmental Services, Gainesville, Florida, 1987); this practice also protects nests from beach-cleaning equipment, if the nests are not moved before cleaning.

At night, nesting turtles are easily disturbed by humans on the beach. Murphy (1985) reported that beach disturbance can cause turtles to shift their nesting beaches, delay egg-laying, and select poor nesting sites. Public education is being used to alleviate the problem.

Experimental Conservation Practices

Headstarting

"Headstarting" is the term used to describe an experimental procedure wherein hatchlings are retained in captivity and reared for at least several months to increase the juvenile population by reducing hatchling mortality. Despite several years of the headstart programs and the development of good husbandry techniques at some facilities, the value of the technique is still debated. Survival of headstart turtles for several years in the wild has been documented, but no nesters of headstart origin have been found. Supporters of headstarting argue that recruits might have been missed, that tags fall off, that there has not been enough time for them to reach adulthood, and that the public-awareness component of having many turtles in tanks for people to see is an important positive result of headstarting. From the research point of view, headstarting has proved valuable in increasing understanding about elements of physiology and behavior of sea turtles (Owens et al., 1982). In 1989, Florida decided to terminate its 30-year-old green turtle and loggerhead headstart program on the grounds that "possible interference with imprinting mechanisms that guide turtles to the nesting beach, imbalance in sex ratios from artificial incubation of eggs, nutritional deficiency from confined maintenance of hatchlings, and behavioral modifications are all potentially serious problems that are cause for concern" (Huff, 1989). In addition, the practice is expensive.

The U.S.-Mexico cooperative headstart program for the Kemp's ridley has been the responsibility of the NMFS Galveston laboratory since 1978. Entry of headstarted turtles into the nesting population has not yet been documented. Wibbels et al. (1989) recommended that headstarting be continued as a research project but that the effort not be expanded. They also suggested that increase in public awareness of the sea turtle situation and the development of strong international collaborative ties between the United States and Mexico were both worthwhile aspects of this particular headstart experiment.

We found no adequate sample of natural hatchling survival against which to judge the success of headstarting. Also, headstarted turtles might be too naive to survive in the wild, and that could undo any positive effects of avoiding the high early mortality in nature.

Before evaluating the headstarting experiment and determining whether the technique should become a conservation practice, one must consider whether four sequential milestones have been reached. In order of achievement, they are growth and survival of headstarted turtles once they are introduced into the wild, nesting of some headstarted turtles on a natural beach, nesting of enough turtles to contribute to the maintenance or recovery of the population, and demonstration that a headstarted turtle is more likely to survive and reproduce than one released as a hatchling. There are still reservations concerning the first milestone in that some released headstarted turtles appear to show maladaptive behavior patterns, such as swimming up to boats in marinas or crawling on beaches (pers. comm., K. Bjorndal, University of Florida, 1989); nevertheless, many recaptures indicate that turtles are feeding and growing in the wild (Manzella et al., 1988). There is no indication of success regarding the remaining milestones in any headstart experiment.

During the 11 years of the experiment, substantial improvements in the protocol have been introduced as new technology and experiential insights have been realized. Specifically, during the early years, male-skewed sex ratios were produced (Shaver et al., 1988), and suboptimal or trawler-occupied release sites occasionally were used. More recently, improved physical-fitness techniques have been developed, improved health-care and nutrition practices have been implemented, new tagging technologies have been adapted to improve the likelihood of identifying headstart turtles after several years in the wild, and the Padre Island artificial imprinting component has been discontinued. In effect, the current experiments are quite different from the original design and offer clearly improved chances for success.

The Kemp's ridley headstart program is a continuing research program that has produced useful information on sea turtle husbandry, behavior, and physiology. However, it is not yet considered to be a long-term management tool in the recovery of endangered sea turtles. It is unlikely that headstarting will ever meet its goal of increased recruitment into the adult populations without a simultaneous reduction in juvenile mortality in the wild based on the analysis of reproductive value by Crouse et al. (1987).

Captive Breeding

Loggerheads, green turtles, and Kemp's ridleys have been raised in captivity from eggs to adults. The same species have laid fertile eggs in captivity. The Cayman Turtle Farm, Ltd., on Grand Cayman Island, has had the most notable success in that regard, rearing both green turtles and Kemp's ridleys from eggs to reproductive adults (Wood and Wood, 1980, 1984). Of the three "experimental conservation" practices most commonly attempted with sea turtles (headstarting, artificial imprinting, and captive breeding), only captive breeding has actually been shown to be successful. Thus, a worst-case alternative strategy to save the sea turtle species in captivity is available, in case they ever disappear from the wild (Owens, 1981). In the case of the Kemp's ridley, retaining captive individuals could serve as a form of genetic insurance, in case a catastrophic event wiped out most of the natural population. The committee emphasizes that this approach would be a method of last resort, and a risky one at best, because captive animals in an aquarium or zoo retain only a portion of the genetic material of their species in the wild.

Artificial Imprinting

Carr (1967) discussed the theory of natal beach olfactory imprinting as it might apply in marine turtles. An extension of this theory is the experimental application of artificial imprinting, in which it has been assumed that, if hatchlings do imprint, the imprinting cues can be altered to a new beach by relocating the eggs to the new beach for their incubation and hatching, emergence, and movement of hatchlings into the ocean. In this process, it has been hoped that new nesting sites could be created or old ones restored. Whereas the entire process is not well understood or proven, some limited evidence suggests that it does occur (Grassman et al., 1984) but that it might be more complicated than initially thought (Owens and Morris, 1985; Grassman and Owens, 1987). Owens et al. (1982) discuss the implications of artificial imprinting in conservation.

REDUCING ADULT AND SUBADULT MORTALITY ASSOCIATED WITH HUMAN ACTIVITIES

Intentional Harvest of Sea Turtles

The deliberate capture of sea turtles was outlawed in the United States by the progressive inclusion of the various species on the Department of the Interior lists of endangered and threatened species. That action protected domestic populations of sea turtles and their eggs, and also outlawed the commercial importation and sale of all sea turtles and their products.

The extent of breach of the regulations within the United States is obviously difficult to assess. However, although occasional persons reportedly are apprehended with a few hundred turtle eggs gathered or offered for sale, the problem does not appear to be serious, compared with the loss of eggs through other causes (e.g., beach erosion) or the loss of immature or mature turtles to incidental capture.

Outside the United States, various laws apply, with various degrees of success. Most nations of the wider Caribbean basin are now parties to the Convention on International Trade in Endangered Species (CITES), which legally bars them from engaging in international commerce in sea turtles and their products. Moreover, turtle eggs receive legal protection in many countries, including Mexico, where sea turtle eggs of all kinds first received legal protection in the Tabla General de Vedas (General Schedule of Closed Seasons), which has been strengthened and extended several times by laws that protect turtles or establish closed seasons for them. A terrestrial reserve and a no-trawling zone have been established in the area critical for Kemp's ridleys in southern Tamaulipas. It is most important that compliance with these laws be strictly enforced. Human harvest of turtle eggs and slaughter of animals continue to be potential problems in Mexico; at the Rancho Nuevo beach, Kemp's ridley eggs must be and are removed to a protected hatchery within hours of their being laid to avoid predation by coyotes or humans.

In the Bahamas, complete protection is given to all life stages of the hawksbill, and eggs and nesting females of all turtle species are protected. There is a closed season on the harvest of all turtles from April 1 until July 31, and minimum-size limits are in effect for green turtles (60 cm SCL) and loggerheads (76 cm SCL) for the rest of the year. Leatherbacks are seen only rarely in the Bahamas and are not taken for food. Kemp's ridleys have not been reported in the Bahamas. Enforcement of regulations, particularly in the more remote islands, is difficult.

In most of the Caribbean, sea turtles have at least some legal protection at some times, although enforcement is often lacking. Costa Rica initiated a legal quota of 1,800 green turtles per year in 1983, but lowering of the quota is being considered. Although the committee is concerned about the effects of intentional harvest outside the United States on sea turtle populations, it has not been able to quantify the extent of the problem.

Incidental Capture of Sea Turtles

Shrimp-Fishing Operations

Various fisheries in U.S. waters have an impact on sea turtles. Deaths related to some fisheries have been well documented (see Chapter 6), in particular the bottom-trawl fisheries of the Gulf of Mexico and the south Atlantic states. Several management tools are available for reducing the impact and might be used in combination for optimal management.

Regulation of fisheries typically uses one or more of the following approaches: limiting the number of individuals that may be captured (zero, in the case of endangered species), limiting the amount of fishing effort with a particular gear type, and controlling the efficiency of a particular gear type. Effort is the amount of time a particular gear type is used; efficiency can be thought of as a measure of a particular gear type's tendency to capture or kill organisms of a target species.

Controlling Trawling Effort

Limitations on trawl-fishing effort can span a continuum from sweeping bans on the use of trawl gear to focused time and area closures. Legal authority for any of those measures can be found in the Endangered Species Act.

An absolute ban on trawling in waters where encounters with sea turtles occur has advantages for eliminating trawl-related turtle deaths and for ease of enforcement. But its socioeconomic impacts are equally clear, constituting an impressive array of disadvantages.

A less extreme approach is to implement time and area closures to reduce the impact of trawling as turtles occupy an area or are especially vulnerable to trawl-related death; this approach has already been used off Rancho Nuevo in Mexico during the nesting season of Kemp's ridleys. "Area" could be defined to include depth zones, as well as more conventional geographic regions. The greatest disadvantage of time and area closures is that their broad application on fine time/space scales might require more and better information than is available on the distribution of sea turtles (see Chapter 4). If such information became available, enforcing such closures might still be challenging, given the difficulties of tracking numerous fishing vessels and monitoring their activities. Another problem might arise if the times and areas closed were so great or coincided so closely with optimal fishing patterns as to make fishing uneconomical.

In several areas and times of the year, turtles might be sufficiently low in abundance that shrimp fishing could be conducted without the use of tow-time restrictions or turtle excluder devices. One area that should be considered is water deeper than about 27 m in the Gulf of Mexico, where juvenile and adult turtles apparently are only about one-tenth as abundant as in shallower waters (see Chapter 4). Some shrimp fishing occurs at that and greater depths in the gulf. It would be necessary to reevaluate the practice after sea turtle populations began to recover and turtle abundance increased in the gulf, to be certain that any turtles on or near the surface were not captured.

The potential of shrimping in fishing zones and times of the year where damage to turtle populations would be minimal without turtle excluder devices or tow-time restrictions should be examined in detail, initially from existing data bases. Some of the difficulties in devising such a management scheme on a large scale become apparent when one examines the material in Chapter 4 on distribution and Chapter 6 on sources of mortality associated with human activities. First, there is a great deal of overlap in the distribution of sea turtles and fishing effort throughout the year. Second, most measures of turtle abundance are not independent of fishing effort; for example, sea turtle strandings are the result of a complex interaction between sea turtle abundance and shrimp trawling (Chapter 6). Third, aerial surveys, although independent of fishing effort, do not detect the smaller turtles such as Kemp's ridleys and juvenile loggerheads, both of which require protection. Fourth, areas in which turtles are now rare enough at some times of the year not to be caught in trawls might be that way only because populations are severely depleted.

Manipulating Trawl Selectivity and Efficiency

Negative effects of trawling can be reduced by modifying the gear so that it will not capture sea turtles or so that captured turtles can escape from the trawl gear without harm. Such modifications can be used in conjunction with effort limitations, as is called for in existing regulations.

The various TEDs approved by NMFS are all designed to be installed in shrimp-trawl gear with the purpose of releasing sea turtles and other large objects from the net without releasing shrimp. Such a separation is mechanically feasible, because turtles are so much larger than shrimp. To some degree, the effectiveness of separation also relies on differences in behavior of various species trapped in a trawl.

Because of the relatively high concentrations of sea turtles in the ocean waters offshore of the Cape Canaveral region of Florida's east coast, NMFS has used this area to assess the effectiveness of alternatively designed TEDs. If a particular TED can be shown to exclude at least 97% of the sea turtles otherwise captured and retained in a control trawl without a TED, that TED is certified by NMFS as an approved TED that meets the requirements of the regulation. By November 1989, six different TEDs had met the minimal criterion for excluding sea turtles and have been approved: the NMFS TED, the Georgia jumper TED, the Cameron TED, the Matagorda TED, the Morrison soft TED, and the Parrish soft TED. (See Appendix C for diagrams of approved TEDs.)

Although each approved TED effectively excludes sea turtles, a TED need not be effective in retaining shrimp to be approved. Furthermore, NMFS is under no legal obligation to assess the effectiveness of each approved TED in retaining shrimp. Nevertheless, a TED is of no value to the shrimp fishery if it excludes too high a percentage of the shrimp that would otherwise have been caught. In announcing the June 27, 1987, regulation that required use of TEDs by shrimp trawlers in most shrimping grounds during most of the shrimping season in the Southeast, NMFS referred to its own test data on the effectiveness of the NMFS TED in excluding sea turtles in offshore waters around Cape Canaveral and retaining shrimp under commercial shrimp trawling in most southeastern states (*Federal Register*, Vol. 52, No. 124, pp. 24244-24262). During TED tests for excluding turtles at Canaveral, low concentrations of algae, debris, and shrimp were encountered. Shrimp loss was very low, averaging a statistically nonsignificant 4% of total numbers and total poundage. Other TEDs might be less effective in retaining shrimp under the same conditions; the (modified) Parrish soft TED, for example, was approved, because it met the minimal standards for excluding sea turtles, but it lost 80% of the shrimp catch as compared with the control (*Federal Register*, Vol. 53, No. 170, pp. 33820-33821).

Numerous tests of the effectiveness of different TEDs in retaining shrimp and of modifications of TED assembly and installation have now been conducted by NMFS, Sea Grant researchers, and state fisheries agencies. It is clear that shrimping efficiency of trawls equipped with TEDs is highly variable based on differences in the specific TED, location, and shrimping conditions. For example, Report No. 7 of the NMFS Observer Program for TEDs documents a range of effects of TED use on shrimp catch of –45% to +38% by weight. The average effect of TED use varied across test regions from –2% to –27%, with most region-specific means between –4% and –15% (pers. comm., E. Klima, NMFS, 1989).

Understanding the variation in shrimping efficiency of TED-equipped trawls is necessary for an evaluation of whether TEDs constitute a solution to the dilemma of how to exclude turtles from trawl nets without economically affecting shrimp catch. Some of the variance in shrimp catch is a consequence of differing performance characteristics of the different TEDs. For example, under identical conditions in offshore waters with little algal debris, Holland (1989) demonstrated a statistically nonsignificant loss of only 4% of shrimp by weight with a Georgia jumper TED (3% with a 4-inch grid and 5% with a 2.3-inch grid), compared with a 54% loss with a Parrish soft TED and a 27% loss with a Morrison soft TED. Hard TEDs appear generally preferred by shrimp fishermen, especially in areas with little debris, probably because of their superior shrimp-retention characteristics (pers. comm., D. Harrington, University of Georgia Sea Grant, 1989).

Another important source of variation in the shrimp fishing performance of TEDs is the variability in concentration of debris in the bottom waters and on the bottom. In areas with abundant debris, it is reasonable to expect a TED to collect some debris, and that will alter its performance. For example, as plant detritus and other debris collect against the bars of an NMFS TED, the shrimp that would ordinarily pass through the gaps unhindered are instead likely to be deflected toward the exit door. Debris is also likely to clog the exit door and prop it partially open, thereby contributing to the loss of shrimp. The shrimp retention rates of other types of TEDs are also affected by debris. Graham (1987), working on the Texas coast, demonstrated that shrimp catch with a Morrison soft TED was reduced by 16% in the presence of abundant sticks and bottom debris that had been naturally deposited from riverine runoff after a rainstorm (Table 7-1). Controlled tests of the Georgia jumper TED recently completed in inshore waters of Core Sound, North Carolina, under normal inshore conditions of abundant seagrass debris and tunicate clumps, demonstrated shrimp losses of 26% by weight (one-tailed paired t test, $n = 15$; $p = 0.016$), compared with control trawls without a TED (pers. comm., C.H. Peterson, University of North Carolina, Oct. 1989; 1990). In this nighttime test, the average shrimp catches were 8.1 pounds with the TED and 10.9 pounds without. In the North Carolina study, only 30-minute tows were used. Such a short tow time would be expected to underestimate the shrimp loss experienced in a fishery, where tow times of 60 to 90 minutes are typical, because a TED-equipped trawl probably fishes efficiently at first and then loses shrimp after becoming clogged. Fouling and clogging of TEDs are likely to occur in areas with high concentrations of plants and other debris near and on the bottom, such as seagrasses, sargassum, various macroalgae, plastic bags, tunicates and other large epibenthic invertebrates, tree branches, lost fishing gear, and debris from oil and gas exploration. Those conditions are probably characteristic of many inshore waters (sounds, estuaries, lagoons, and coastal embayments), which generally lie closer to seagrass or benthic algal beds and closer to sources of human-discarded debris, but they can occur at times in any locality, even in offshore waters.

Under some conditions, use of a TED might improve shrimp catch. The basic TED design is a modification of a "cannonball shooter," a device first developed to eject large cannonball jellyfish (Stomalophus meleagris). When those jellyfish are abundant, tow times must be drastically reduced, because the weight capacity of the net is quickly reached. That greatly lowers the proportion of time that nets are actually fishing and thus reduces shrimp catch. Under such conditions, use of an effective TED enhances shrimp catch.

Not all the observed variation in shrimping efficiency of TEDs is explained by differences in TED design, concentration of debris near and on the bottom, or size of the finfish bycatch. But it is clear that conclusions based on specific TED design used under a particular set of conditions cannot be extrapolated to all TEDs and all conditions. NMFS has tested alternative TEDs off Cape Canaveral, an area with high densities of sea turtles and low concentrations of debris and shrimp. However, even the results from that site cannot be extrapolated to all conditions. Further tests are needed to identify other covariates that contribute to variation in shrimping efficiency and perhaps even in the effectiveness of sea turtle release. For example, bottom topography and sea roughness are likely to affect TED performance in shrimping. Insufficient data are available to evaluate whether TEDs clogged by algal and other debris continue to release sea turtles effectively and whether TEDs eject small sea turtles as effectively as medium or large turtles.

One further difficulty is related to the dynamic nature of trawl gear. A trawl is a flexible bag of netting whose shape (and therefore function) is determined by hydrodynamic forces and friction and whose geometry and performance are highly variable. The flexibility of trawl gear is one of the challenges constantly faced by gear designers and fishermen, and in fact it is exploited by successful fishermen to "tune" their gear to accommodate changes in fishing conditions or in their own fishing needs. A TED will inevitably affect the balance of forces that determine the geometry and function of a trawl. Likewise, changes in the geometry of the trawl due to the highly variable nature of the fishing environment will affect the performance of a TED, especially if the TED is of a soft design. Current procedures for certifying TEDs make no accommodation for those properties of gear and afford the well-intentioned, competent fisherman little scope for modifying installed TEDs, even though some "tuning" might be needed to get a TED to function well without unacceptably reducing shrimp-catching efficiency. For TEDs to work in a commercial fishery, the fishermen must be motivated to make them work and then

TABLE 7-1 Effect of debris on the relative efficiency of shrimp capture by a trawl equipped with a Morrison soft TED.

Clean Tows				Fouled Tows			
N	Shrimp Catch (lb) No TED	Shrimp Catch (lb) with TED	Average Difference (±SE)	N	Shrimp Catch (lb) No TED	Shrimp Catch (lb) with TED	Average Difference (±SE)
			Charlene M.				
24	49.5	47.6	–1.9* (±1.1) [-3.8%]	18	58.2	49.1	–9.1† (±2.2) [–16.0]
			Sea Tiger				
20	27.6	27.6	–0.1ns (±0.5) [–0]	10	54.0	45.5	–8.5* (±4.0) [–16.0]

NOTE: Data come from two sets of TED test cruises in September 1987, run by G. Graham, one on the *Charlene M.*, the other on the *Sea Tiger.* Data come from contrasts of paired parallel trawls, one with and the other without a TED.

Each pair of trawl tows was classified as fouled with debris (sticks, seagrass, etc.) or clean on the basis of contents in the tail bags. N equals the number of pairs of tows. Numbers in brackets are the percentage difference in shrimp catch between the control net (net without TED) and the net with a TED.

For a one-tailed paired t test, ns = $p > 0.05$; * = $p < 0.05$; † = $p < 0.0005$.

given the latitude to experiment with them until they do. TED use can be legislated, but effectiveness requires the cooperation of fishermen; without it, a process of proliferating regulation of gear design and fishing tactics will be initiated.

The committee also notes that considerable technical resources are available that could be applied to the improvement of TED design.

Tow-Time Limits

Available data (Henwood and Stuntz, 1987) show that the average rate of mortality of sea turtles captured in trawls is reduced to a negligible point (less than 1%) as tow time is reduced to 60 minutes or less (Figure 6-3). Total tow times (defined as the actual bottom fishing times) of 90 minutes cause substantial mortality of captured sea turtles. Assuming that the proximate cause of sea turtle mortality in shrimp trawls is drowning, one should reasonably question on physiological grounds whether the information on drowning published by Henwood and Stuntz (1987) is sufficiently partitioned to develop management regulations that are adequate to protect sea turtles. Because respiratory demand for oxygen is expected to vary with turtle species, body size, time of day, and temperature, we obtained the NMFS data set used by Henwood and Stuntz (1987) and used it to study the following contrasts: loggerheads versus Kemp's ridleys, large vs. small animals, nighttime vs. daytime capture, and winter vs. summer trawling. We compiled the data by 10-minute tow-time intervals to allow variance to be more readily observed.

The results imply substantial seasonal differences (but less variation in other factors) within the data set that are of significance to management of the shrimp-trawling fishery. First, there is no obvious large difference between loggerheads and Kemp's ridleys in the curve that relates mortality in trawls to tow time: tow times of about 60 minutes or less appear to cause negligible mortality on the average in both species. Second, we could detect no large difference between the two size classes of sea turtles: a tow-time limit of about 60 minutes produced negligible mortality in each. Third, there is a suggestion of only a small difference between the daytime and nighttime curves: for tow times less than about 60 minutes, turtle mortality is negligible. Fourth, there is a difference between summer and winter curves (Figure 7-1, top): in the summer, when respiratory demands are presumably greater at the higher water temperatures, a tow-time limit of about 40 minutes appears necessary to ensure negligible mortality of captured sea turtles; in the winter a tow-time limitation of about 90 minutes has equal effectiveness.

Two additional questions need to be addressed before one can have complete confidence in the effectiveness of a tow-time regulation. First, evaluation of available data or collection of new information is needed to assess how frequently shrimp trawlers are concentrated so that multiple captures of the same turtle occur without adequate recovery time and thus lead to even higher mortality. Second, the paper by Henwood and Stuntz (1987) and the committee's initial breakdown of their data assumed that comatose sea turtles recover. As indicated in Chapter 6, there is reason to believe that significant numbers of comatose turtles die (pers. comm., P. Lutz, University of Miami, 1989). We have also recalculated the mortalities, including the comatose with the dead. Until we know what fraction of comatose turtles actually survive, the tow-time limitation would require adjustment of tow-time limits downward to keep expected sea turtle mortality under 3% of all turtles captured. In Figures 6-3 and 7-1 (bottom), we present data on how numbers of dead plus

FIGURE 7-1 Relation between percentage of dead and comatose sea turtles (mostly loggerheads) in summer versus winter as a function of tow time of trawls (data from Figure 6-3, broken down by season). Top: dead turtles; bottom: dead plus comatose turtles. Total numbers of turtles captured: winter—2,490; summer—1,907. Compiled by the committee from raw data provided by NMFS, which were the basis for Henwood and Stuntz's (1987) calculations.

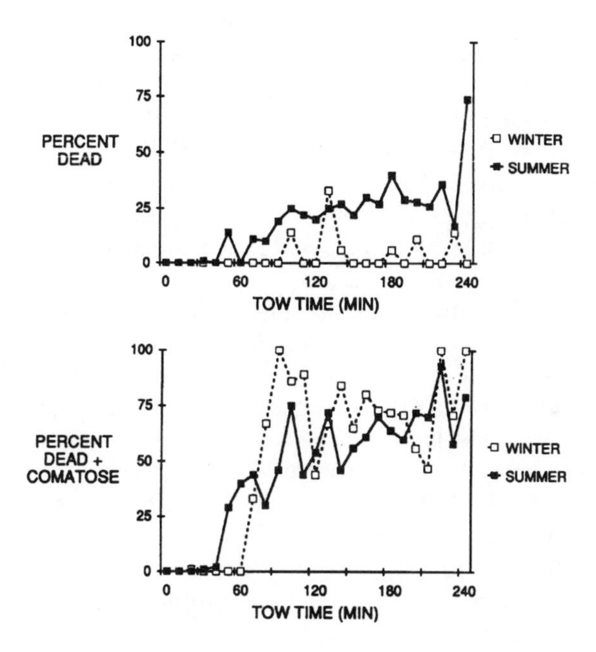

comatose turtles vary with tow time as an upper limit on mortality. Those numbers are far larger than the numbers of dead alone and, if they reflected true mortality of sea turtles, would suggest a need for further reduction in tow times to protect sea turtles from drowning. Specifically, winter tows might need to be restricted to 60 minutes or less, instead of less than 90 minutes, whereas the 40-minute restriction in the summer seems sufficient.

Tow-time limitation could be as effective as TEDs in reducing the mortality of sea turtles in shrimp trawls, but might be extremely complex as a management option, because of differences in seasons and locations. The regulations would need to vary with season, to allow the most efficient shrimping while still protecting the turtles. However, the brevity of acceptable tow times results in a cost to shrimpers, because their nets are fishing for a smaller fraction of the day. Given the relatively small shrimp losses demonstrated in offshore shrimping with the most efficient TED (e.g., Holland, 1989; pers. comm., E. Klima, NMFS, 1989), it seems likely that shrimpers fishing in the offshore waters would catch more shrimp by using a TED than by restricting tow times to 40 minutes in the summer and 60 minutes in the winter. In inshore waters and in other situations with concentrated plant and other debris, trawl times are necessarily limited, usually to less than 90 min, by the accumulated weight of debris (including finfish bycatch) in the trawl. Under those conditions, imposition of a tow-time restriction adds little cost to what nature imposes. Those are also conditions under which TEDs fail to retain a high fraction of the shrimp entering a trawl and, more important, under which the effectiveness of many TED designs in releasing turtles might be compromised. Thus, tow-time limitation in coastal waters of estuaries, sounds, lagoons, and embayments constitutes a sensible management tool in some areas. The 1987 NMFS regulation recognized a 90-minute tow time as an alternate to TED use in inshore waters only.

A major concern regarding the use of tow-time limitation as a management tool is how it can be enforced. The problem has not been solved, but new technology could be directed toward engineering a device to record submergence time. Enforcing proper use of TEDs is also a major concern, because TEDs can be readily disabled by altering the tension of spring cords or tying them in a fashion virtually undetectable by inspectors. All this suggests a need for evaluation of the effectiveness of regulations of the shrimp-trawl fishery.

The physiology of prolonged forced submergence needs further study, to allow for the complete evaluation of the use of tow-time limits in trawl fisheries. Even a "normal" appearing turtle that has survived 60 minutes of compression and forced submergence might have lung, heart, or other vital organ damage (Manzella et al., 1988). How enforced submergence affects sea turtle physiology as a function of season, water temperature, turtle species and size, time of day, and history of previous enforced submergence is not well known. For example, is recapture more likely in turtles that have just been released from another trawl? The impact of multiple recaptures on sea turtle survival might lessen the effectiveness of reduced tow-time regulations in saving turtles.

Finally, there is the persistent question of multiple physiological stresses that might act on one another. The potential needs further evaluation, although we do not believe that a shortage of knowledge affects the recommended conservation measures identified in this report. Wolke (1989) believed that the health of many of the dead sea turtles that he necropsied might have been compromised by parasites. After a decade of observer programs in which fresh carcasses have been available from trawls, it is surprising that so few necropsies have been done on fresh wild carcasses. Much more could be learned about the physiological condition of sea turtle populations and the possible interactive effects of multiple stresses, if more professional necropsies are performed on fresh carcasses.

Other Commercial Fishing Activities

Various commercial fishing activities besides shrimp trawling kill sea turtles. In some cases, turtle deaths have been observed (Chapter 6); in other cases, no observations were made, but the nature of the fishery or other considerations suggested at least a potential for harmful encounters.

Some closures have been implemented for fisheries other than shrimp fisheries. Ocean gill nets set to capture sturgeon are now prohibited by state fisheries regulations in both North Carolina and South Carolina, reducing the incidence of sea turtle mortality apparently associated with this activity (Murphy and Hopkins-Murphy, 1989). For set net fisheries, change in mesh size could reduce entanglement of sea turtles.

Dredging, Boat Collisions, and Oil-Rig Removal

Dredging, boat collisions, and oil-rig removal were each estimated to kill from 50 to 500 loggerheads and five to 50 Kemp's ridleys a year, if mitigation or conservation measures were not in place.

Dredging

When it was first noted that large numbers of turtles were being taken by dredging within the Canaveral Channel (Joyce, 1982), NMFS and the Jacksonville District of the U.S. Army Corps of Engineers took immediate action to reduce the problem, including the relocation of 1,250 loggerheads from the Canaveral Entrance Channel to areas offshore for the remainder of the dredging operation (Joyce, 1982). The relocation effort proved to be less than successful: many of the displaced animals returned to the channel in an unacceptably short period. However, another relocation effort in December 1989 and January 1990 in the Cape Canaveral Channel was successful; no relocated animal was recaptured in the channel (pers. comm., A. Bolten, University of Florida, 1989).

Through the mechanism of Section 7 consultations provided by the Endangered Species Act of 1973, a Sea Turtle Dredging Task Force was created in 1981 to respond to concerns by NMFS about the unacceptably large numbers of sea turtles taken during 1980. Members of the task force included representatives of the Army Corps of Engineers, NMFS, the U.S. Fish and Wildlife Service, the Florida Department of National Resources, and the Navy (Studt, 1987). A number of continuing actions have been initiated by the task force to document and mitigate sea turtle losses:

- *Initiation of an observer program with on-board biologists to document the take of sea turtles by dredges, including the modification of gear to screen discharge ports for the presence of sea turtle parts.* The observer program was initiated at the Port Canaveral Entrance Channel in 1980 (Joyce, 1982) and at the St. Mary's Entrance Channel in 1987 (Slay and Richardson, 1988; Richardson, 1990). The observer program will be expanded to additional harbor-entrance dredging operations along the Eastern Seaboard, the Gulf of Mexico, and Puerto Rico, wherever sea turtles are known to occur and as opportunity permits (pers. comm., T. Henwood, NMFS, 1989).
- *Investigation of the configuration and relative threat to sea turtles of various types of dredges and dredge dragheads.* Sea turtle take has been associated primarily with hopper dredges used for offshore channel dredging. Hydraulic cutterhead dredges and bucket dredges, used primarily for inshore work, do not appear to affect sea turtles to a significant degree. Relative to hopper dredges, investigations in 1981 and 1982 identified the California type of draghead as the least damaging (Joyce, 1982) and the gear of choice for Port Canaveral. However, the take of sea turtles with the California draghead in the Port Canaveral Entrance Channel since 1980 has been found unacceptable by NMFS (pers. comm., T. Henwood, NMFS, 1989), so alternative dredging methods and gear types are now being sought for this channel.
- *Design and test modifications of hopper-dredge dragheads.* Various deflector systems have been tested, with minimal success because of the powerful suction force of the intake water and because of the destructive mechanical forces applied to the deflector apparatus on

the channel bottom (Studt, 1987). Efforts continue to develop a functional deflector for dragheads used in the Port Canaveral Entrance Channel during maintenance dredging (pers. comm., T. Henwood, NMFS, 1989).

- *Investigation of various sensory stimuli to repel turtles from the channel to be dredged or from the vicinity of the dredge.* Investigations have not had results that can be applied to mitigation of turtle take by hopper dredges (Studt, 1987). Air guns used in seismic exploration did not deter sea turtles at the Turkey Point, Florida, power plant (pers. comm., J. O'Hara, Environmental and Chemical Sciences, 1989). It is unknown what further research efforts might be attempted in this area.
- *Radio-tracking studies of sea turtles in the navigation channels.* Nineteen loggerheads were tracked in the Port Canaveral Entrance Channel in 1982 (Nelson et al., 1987). Valuable behavioral information was obtained that could be used for censuses, such as the proportion of time spent on the surface and the number of surfacings per hour by an average turtle. Movements of the turtles in the channel and between adjacent habitats proved unpredictable and did not lead to suggestions for mitigation.
- *Determination of the frequency and distribution of sea turtles in key navigation channels of Florida's coast and elsewhere.* Several censuses of sea turtle populations in the Port Canaveral Entrance Channel have been conducted since 1980 (Henwood, 1987) and are continuing. Early results indicated that sea turtles were present in considerable numbers at all times of the year, but in the lowest numbers during September, October, and November (Studt, 1987). Dredging at Canaveral was then restricted to that 3-month period. Seasonal dredging is considered the most important available mitigation measure.

The committee did not have time to analyze the recent Army Corps of Engineers report (Dickerson and Nelson, 1990), but we note that many of its suggested studies are similar to those in the present report.

Collisions with Boats

Estimates of mortality from collisions with boats are uncertain, because the assessment of wounds on stranded animals usually cannot determine whether the turtles were hit before or after they were dead and floating in the ocean. Wounds should be photographed and measured to be certain of their origin. In addition, there are no estimates of collisions in inside waters.

Because 50-500 loggerheads and 5-50 Kemp's ridleys might be killed each year (Table 6-2), judging by the incidence of wounds on stranded animals, a better assessment is needed than is provided by the stranding network. If geographic areas of critical concern are found, methods like those imposed for protection of manatees from boat collisions should be implemented in selected waters off nesting beaches. Distributing information on the problem to boat owners could be helpful, but, because human-turtle interactions are so widely dispersed, substantial reductions in mortality are unlikely.

Oil-Rig Removal

Sea turtle species and turtles in different life stages within species often segregate by habitat preference. The deployment of underwater structures (oil-platform tripods, towers, anchors, sediment-control devices, ocean cables, and the like) and other marine activities (mining and drilling) might promote formation of local concentrations of sea turtles in unpredictable ways. The feasibility of removing turtles from the vicinity of all planned explosive detonations must be investigated.

Power Plant Entrainment

Power plants have minimal influence on sea turtles, killing perhaps 5-50 loggerheads and Kemp's ridleys each year. The measures described in Chapter 6 that are now in place for the St. Lucie No. 2 plant seem adequate and should be continued. Further evaluation and intake system modifications might eventually be necessary at other plants, where larger numbers of turtles could be entrained and killed as populations increase in the future.

Ingestion of Plastics, Debris, and Toxic Substances

There is ample evidence that sea turtles ingest plastics and other indigestible materials of human origin (see Chapter 6). For example, Plotkin and Amos (1988) found plastics and other debris of human origin in 46% of 76 carcasses necropsied on the Texas coast. Further documentation is needed of the extent of the problem, particularly the mortality rate associated with ingestion, the physiological response of the animal to ingested materials of different types and particle sizes, and the behavioral response of turtles to oceanic debris.

All carcasses should be checked for the presence of ingested plastics.

Drift lines of sargassum and other materials at sea should be checked for the presence and characteristics of plastics and for the occurrence of turtles ingesting plastics. Materials found at sea and on beaches should be checked for evidence of feeding by turtles. The tendencies of turtles to ingest plastic debris of various types, particle sizes, and colors should be checked under controlled conditions. The ability of selected research animals to pass ingested plastics of particular types and particle sizes without physiological damage should be determined. The wording of MARPOL and other ocean and coastline dumping regulations should be examined for applicability to the problem of plastics ingestion by sea turtles, particularly statements related to maximal allowable particle size of shredded materials discarded overboard, and the implementation effectiveness of dumping regulations should be investigated.

Additional information is needed on the reaction of sea turtles to petroleum ingestion, fouling, and toxicity. Fritts and McGehee (1981) found that sea turtle eggs contaminated with fresh crude oil, as might occur after an oil spill, yielded a lower hatch rate and a higher percentage of deformities. They did not, however, investigate the effects of floating oil on the behavior of animals in the water—courting, mating, feeding, and the like. Lutz and Lutcavage (1989) exposed young loggerheads to very brief contact with crude oil and found reduced hematocrit measurements, modified behavior, and alterations in skin epithelium. They felt that more work was required to document fully the impacts of crude oil. Sea turtles ingest tar (which is chemically passive) and oil droplets (chemically active) that they appear to mistake for food particles (Witham, 1978; Lutz and Lutcavage, 1989). Such materials are abundant in the pelagic environment, particularly in drift lines, so a better understanding of the physiology of ingestion of these materials as they pass through the intestine is needed.

Drift lines and samples of plankton and ocean-surface particles should be checked for the presence and characteristics of crude-oil derivatives. The presence and effect of tar should be documented at all necropsy opportunities, including correlation of particle size and abundance with size and condition of turtle by species. Moribund animals should be looked for in the vicinity of oil spills and concentrations of petroleum particles, especially in the Gulf of Mexico where Kemp's ridleys are found. The physiological response of selected animals of different sizes and species to ingestion of floating tar particles passing through the intestine of the research animals should be investigated. And the wording, implementation, and enforcement of national oil-spill regulations and international protocol should be checked for responsiveness to the needs of sea turtle conservation.

EDUCATION AND TECHNOLOGY TRANSFER

Education with respect to beach management, reduction of human-associated mortality of eggs and hatchlings, and the implementation of technology is important for the conservation of sea turtles. We present here some information on education with respect to beach management, but also focus attention on the implementation of TED technology in the shrimping industry. It will always be difficult to implement an important conservation measure if it is viewed as an economic liability to the user; education should promote the implementation of new useful measures.

Education

One of the easiest ways to implement good beach management is to inform and educate the public. Beach residents conducting turtle projects often advise tourists on what they can do to minimize disturbance to nesting turtles, protect nests, and rescue disoriented hatchlings. Similarly, state, federal, and local parks that conduct beach walks provide information to visitors. Beaches are also posted with signs informing people of the laws that protect sea turtles and providing a local or hotline number for reporting violations.

A wide variety of materials are available (e.g., children's coloring books, posters, slide-tape programs, brochures, and fact sheets) from the Center for Marine Conservation, Florida Power and Light Co., NMFS, FWS, and environmental groups.

TED Technology Transfer

One of the responsibilities of NMFS is to monitor and enforce provisions of the Endangered Species Act. NMFS Southeast Regional Office programs include regulation development, recovery planning and implementation, information dissemination, TED certification, and permit administration. Research activities involve TED technology transfer (providing assistance to industry and evaluating TEDs for certification), TED economic evaluation (tow-time observer program), TED-regulation evaluation (systematic strandings), sea turtle biology and ecology, and Kemp's ridley headstarting (Oravetz, 1989).

NMFS sea turtle program funding since 1977 has averaged $890,400 a year, starting at $250,000 in 1977 and with a high of $1,150,000 in 1982. Sources of additional funding have been Marine Fisheries Initiatives (MARFIN), the Entanglement Network, the Army Corps of Engineers, the

Minerals Management Service, the shrimp industry, the National Sea Grant College Program, FWS, various regional power plants, and Saltonstall-Kennedy (S-K) funds. Current contracts with the Gulf and South Atlantic Fisheries Development Foundation (GSAFDF) total $862,000 (MARFIN and S-K) and are being used to coordinate industry and Sea Grant efforts in TED testing, development, and technology transfer.

The responsibilities of at least 21 NMFS employees include some aspect of TEDs or educating the public about sea turtle conservation. The NMFS laboratory at Pascagoula, Mississippi, maintains three full-time TED-gear specialists employed to help fishermen and net shops build and use TEDs correctly. Four additional gear specialists at the same laboratory have conducted at-sea TED demonstrations and given many presentations (pers. comm., C. Oravetz, NMFS, 1989).

Between 1981 and 1986, NMFS conducted programs to encourage the voluntary use of TEDs by shrimp fishermen. The program involved workshops and TED demonstrations for shrimp fishermen; commercial fisheries associations; reporters; and Sea Grant, state and university personnel. Numerous presentations of TEDs were given at commercial fisheries association meetings and conventions and on the docks and on decks of shrimp trawlers. NMFS has provided 300-400 free TEDs to fishermen to test and use. Slide programs, video tapes, brochures, and instructional materials on TEDs and on sea turtles and their conservation and management were also developed and made available (pers. comm., C. Oravetz, NMFS, 1989). NMFS changed the meaning of "TED" to "trawling efficiency device" in 1983, in hopes of making the gear more palatable to shrimp fishermen. Few fishermen responded; by 1986, less than 3% of active trawlers had used TEDs (*Federal Register,* Vol. 52, No. 124, pp. 24244-24262).

In addition to its inhouse efforts, NMFS cooperates regularly with fishing industry associations and with Sea Grant, state agency, and environmental groups. Sea Grant has been called on to play a major role in education, particularly in TED-technology transfer (pers. comm., C. Oravetz, NMFS, 1989). NMFS contracted with all Sea Grant Marine Advisory and Extension programs in the Southeast to distribute TEDs to fishermen, experiment with TEDs, train fishermen to use TEDS, and generally keep fishermen informed of the ever-changing status of TED regulations.

Sea Grant Marine Advisory and Extension Service personnel throughout the Southeast were asked to play an integral part in assisting NMFS with TED testing and technology in their own states. Since 1981, numerous workshops and demonstrations have been conducted for fishermen and net-makers. Newsletters continually apprise fishermen of planned workshops, demonstrations, and public hearings and of the latest updates of TED regulations. Educational materials have been developed and distributed for some of the certified TEDs, providing diagrams and detailed instructions for proper placement and use of the TEDs in a trawl. The R/V Bulldog of the University of Georgia is often used to test and compare TEDs against standard (non-TED) nets and other TED models near Cape Canaveral, as a precursor to certification. The cruises often include industry representatives, as well as university, Sea Grant, and NMFS personnel. Sea Grant personnel have distributed free TEDs, provided lists of TED manufacturers, and informed fishermen on how to get reimbursement for purchased TEDS from state assistance programs. Some of the Sea Grant personnel have worked closely with net-makers and commercial fishermen to design new TEDs that will reduce shrimp loss, but still allowing turtles to escape.

An annotated chronological list of NMFS education efforts centering on shrimp fishermen and TEDS is found in Appendix G, and documentation of similar efforts by Sea Grant personnel is listed in Appendix H. Those appendices provide a detailed overview of the educational programs and endeavors of the agencies by region and year. A list of educational materials on TEDs is found in Appendix I.

CONSERVATION EFFORTS IN OTHER JURISDICTIONS

Mexico

Measures taken by Mexico to conserve sea turtles have involved academics, nongovernmental private groups, and governmental agencies. Since 1966, government turtle biologists, fisheries inspectors, and armed Mexican marines have maintained a presence at Rancho Nuevo to protect nesting Kemp's ridleys, their nests, and eggs. Shrimp trawling offshore of Rancho Nuevo is restricted. Beach patrols on the gulf and Pacific coasts, although somewhat spotty, provide some protection against poaching of adults and eggs for all sea turtles.

The Instituto Nacional de Pesca, in consultation with FWS, has experimented with TEDs, but to date, TEDs have not been adopted as a Mexican governmental regulation in shrimp trawling.

Olive ridleys are legally harvested under a quota arrangement on the Mexican Pacific coast. By restricting harvest until after the turtles have nested, the government has afforded some protection to these turtles, but a drastic lowering of the quota or cessation of harvest should be seriously considered. Protection of all sea turtle adults and eggs, whether for food, export, or crafts, should be a prime action by the Mexican government.

Other International Concerns

Worldwide efforts to conserve sea turtles have been identified by Bjorndal (1981) and Groombridge (1982). Throughout most of their ranges, sea turtles continue to be under threats of decline from human activities, and are the continuing subject of intense study, regulation, and international action and concern. This is well demonstrated by recent statements and action at the World Herpetological Congress in Canterbury, England (September 1989), the Convention on International Trade in Endangered Species (November 1989), and the International Union for the Conservation of Nature Marine Turtle Specialist Group.

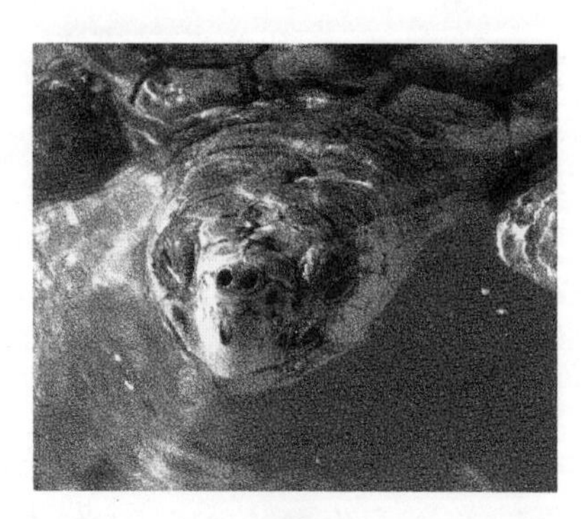

8 Conclusions and Recommendations

CONCLUSIONS

The committee has analyzed available data on the five sea turtle species found in U.S. territorial waters to ascertain current population trends. The most important data were the numbers of nests or nesting females on the nesting beaches, but other useful data were the incidences of turtle strandings and numbers of adults sighted at sea in the course of aerial surveys.

Population trends of several species were evident especially from counts of nests and nesting females. The Kemp's ridleys on the nesting beach at Rancho Nuevo, Mexico, have declined to about 1% of their abundance in 1947; since 1978, the number of nesting females has declined from about 700 to an estimated low of only 350 in 1989. This species is obviously the most critically endangered of all the sea turtle species. Counts of loggerheads nesting on various beaches of the southeastern coasts vary with latitude: numbers of nesting loggerheads on South Carolina and Georgia beaches are declining; but in two study areas in Florida, the numbers seem to be stable in one and appear to be increasing in the other. Green turtles are showing some increased nesting on Hutchinson Island, Florida. Leatherbacks and hawksbills nest too infrequently on southeast beaches for clear-cut trends to be identified.

As judged from stranding data, the most abundant turtles in U.S. coastal waters are loggerheads; Kemp's ridleys and green turtles are less abundant; and leatherbacks and hawksbills are even less common. Aerial surveys designed to count turtles at sea have yielded less precise numbers, because many adult turtles and small juveniles are difficult to identify and count accurately from the air. As a result, aerial surveys have been used only sparingly to assess patterns of sea turtle distribution; their results, however, support the belief that loggerheads are the most abundant species in both inshore and offshore habitats.

Causes of sea turtle deaths, and especially causes associated with declining populations, can be categorized either as natural or related to human activities. Sea turtles have long been harvested for their eggs and meat, for their shells (carved for ornaments), and for their skin (leather) and various body parts (oil-based derivatives). Overexploitation of green turtles for a turtle cannery industry in the Gulf of Mexico as early as the late 1800s was largely responsible for their decline in the early 1900s.

A wide range of human activities have been identified as causing sea turtle deaths. Those with effects especially on sea turtle eggs and hatchlings include various beach manipulations (e.g., fortification, deposition of sand, cleaning), the use of artificial lighting, vehicular and human traffic on beaches, and the planting of exotic vegetation. Although mortality data on many of those factors have been found for various sites at various times, the data are generally too sparse and localized for use in quantifying long-term effects on sea turtle species. The committee was better able to quantify human-associated causes of deaths of juvenile and adult sea turtles.

Of all the known factors, by far the most important source of deaths was the incidental capture of turtles (especially loggerheads and Kemp's ridleys) in shrimp trawling. This factor acts on the life stages with the greatest reproductive value for the recovery of sea turtle populations. Strong evidence for the effect of shrimp trawling on turtles came from the following findings:

- The mortality of turtles caught in shrimp trawls increases markedly for tow times greater than 60 minutes.
- Numbers of stranded turtles increase with the opening of shrimp seasons and decrease with the closing of shrimp seasons.
- Loggerhead populations declined in areas where shrimp trawling off their nesting beaches was intense, but did not decline in areas where trawling was not intense.
- The estimated numbers of sea turtles captured by shrimp trawling are large.

Other fishery operations, lost fishing gear, and marine debris are known to kill sea turtles, but the reported deaths are only about 10% of those caused by shrimp trawling. Dredging, entrainment in power-plant intake pipes, collisions with boats, and the effects of petroleum-platform removal all are potentially and locally serious causes of sea turtle deaths. However, these collectively amount to less than 5% of the mortality caused by shrimp trawling. Natural diseases and parasites, and toxic substances can and do kill sea turtles, but their overall effects on sea turtle populations cannot be quantified. Sea turtles commonly ingest a wide variety of plastic substances and petroleum residues that can harm them. Although the ingestion of plastics has been observed, the magnitude of resulting mortality cannot be determined from existing information.

Natural predation on turtles in all life stages, parasitism, diseases, inclement weather, beach erosion and accretion, thermal stress, and high tides are all known to affect populations adversely, especially on the nesting beaches. But the committee concluded that changes in natural sources of mortality are not the causes of observed population declines except in a few localized instances.

Thus, the committee identified population declines in sea turtle populations, and it determined that the most important mortality factor has been the incidental capture of subadult and adult sea turtles in shrimp trawls.

RECOMMENDATIONS

Conservation Measures

The committee considered several options for conserving sea turtles. Rather than recommend specific regulations, the committee has focused on various aspects of sea turtle biology and various sources of mortality. Its recommendations are therefore general enough to permit various management options in some cases. However, it is clear to the committee that at least the Kemp's ridley population is dangerously small and that the species needs increased protection. In addition, loggerheads are declining rapidly in South Carolina and Georgia, and green turtles remain uncommon, although they are beginning to show some evidence of population increase at one site in southern Florida. All of those species need increased protection under the Endangered Species Act and other relevant legislation.

Reducing Incidental Deaths of Juvenile and Adult Sea Turtles

Shrimping Incidental deaths due to shrimping must be reduced. An estimated 5,500-55,000 loggerheads and Kemp's ridleys are killed each year by shrimping activities in U.S. waters. The waters off northern Florida, Georgia, South Carolina, Louisiana, Mississippi, Alabama, and Texas are most critical, but the committee recommends the use of TEDs in bottom trawls at most places and most times of the year from Cape Hatteras to the Texas-Mexico border. At the few places and times where TEDs might be ineffective (e.g., where there is a great deal of debris), alternative conservation measures for shrimp trawling might include tow-time regulations under very specific controls and area and time closures. Restrictions could be relaxed where turtles are and historically have been rare, such as in deeper waters of the gulf.

The committee believes that shrimping *with adequate controls* is compatible with the recovery of turtle populations. Although prohibition of shrimp trawling might be required as a "last-ditch" measure under the Endangered Species Act, appropriate application of existing technology—especially TEDs, innovative new technologies, and other conservation measures have the potential to reduce sea turtle mortality to a level that the populations can tolerate. The committee comments here on some of the available controls.

- **TEDs**. The use of TEDs at all times in all areas could theoretically reduce the capture of sea turtles to 3% of the rate seen without TEDs. However, complicating factors, such as the presence of seagrasses and other debris, reduce the fishing effectiveness of TED-equipped trawls at some times and might even prevent the successful ejection of turtles that enter the trawls. The available data do not show conclusively that significant numbers of sea turtles occupy all waters fished by shrimpers throughout the entire year. However, turtles are present in some areas even where TEDs are not now required. For example, current regulations do not require TED use from northern Florida to Cape Hatteras waters after the end of August, but stranding data and aerial-survey data demonstrate that sea turtles, especially Kemp's ridleys, are in fact still in these waters through December and are suffering trawl-related mortality.
- **Tow-time limits.** Available data suggest that limiting tow durations to 40 minutes in summer and 60 minutes in winter would yield sea turtle survival rates that approximate those required for the approval of a new TED design. Use of tow-time restrictions would avoid the clogging problems experienced when TEDs are used in areas with abundant debris. The 1987 NMFS regulations appropriately incorpo-

rated tow-time limits as an option in inshore waters, where there often is much debris. However, these tow-time regulations need further refinement. Current tow-time limits are too long, if animals described as comatose in prior reports were in fact destined to die. Also, the current tow-time limits are not properly stratified by season to protect turtles adequately in warm seasons.

- **Relaxation of TED use and tow-time regulations at selected locations and times**. TED use and tow-time limits might be selectively applied when and where the probability of capturing sea turtles does not exceed acceptable levels. Available information should be examined for the potential of such fine tuning. Because the overlap between turtle distribution and fishing activities is great, such an approach would have limited applicability, but would perhaps make the regulations less onerous.
- **Limited time/area closure for turtle "hot spots."** Under special circumstances (e.g., in waters adjacent to dense nesting beaches), sea turtle concentrations in defined areas might be temporarily so high, or the turtles so vulnerable, that other conservation measures do not offer adequate protection.

Other Human Activities Sea turtle deaths incidental to other human activities—such as operation of other fisheries, abandonment of fishing gear, dredging, and oil-rig removal—should be addressed and reduced. Finfish trawls kill some turtles. Groundfish trawls are structurally and operationally similar to shrimp trawls, and their potential effects on sea turtles that encounter them are similar as well. Observer data on rates of sea turtle capture and deaths related to groundfish trawls are not available. The committee recognizes the need for NMFS to assess the effects of gill-net fisheries and the winter groundfish trawl industry on the incidence of turtle capture and mortality. If mortality is substantial, NMFS should consider expanding the regulations designed to protect sea turtles from drowning in trawl nets to include all bottom trawls and set nets, not only shrimp trawls. That would protect sea turtles now at risk because of winter groundfish trawling and the setting of unattended nets, such as pound nets and gill nets.

Research and development should continue, in an effort to reduce further the loss of sea turtles in hopper dredges. Modification of dragheads to exclude turtles during maintenance dredging appears to be feasible, and research on modifications continues. Continuing surveys of population numbers and movements within important, frequently dredged entrance channels will provide more understanding of sea turtle behavior that will be applied to improving management designs. Turtles should be relocated away from dredging operations when necessary.

The observer program is essential to measure the success of efforts to mitigate the loss of sea turtles in hopper dredges. Finding turtle carcasses or remains in the hopper sediments is difficult, so the on-board observer program should be continued. Although some preliminary work has been done, hopper dredges must be monitored, especially where sea turtles might be involved.

Other sources of turtle mortality should be reduced. Marine debris and pollutants can kill turtles that ingest them. MARPOL and other programs to reduce marine pollution are in place, and some have recently been strengthened. Sea turtles are affected to some degree by explosions associated with the building and demolition of marine structures, especially those related to the oil and gas industry. MMS and NMFS programs are under way to reduce these potential impacts.

Reducing Directed Harvest of Sea Turtles

Directed harvest of all sea turtle species in U.S. waters should continue to be prohibited. Because of the transnational migration and distribution of sea turtles, population recovery would be substantially improved if all directed harvests of sea turtles were eliminated in other countries as well.

Reproduction-Related Protection

Critical nesting areas, nesting activities, and early life stages (eggs and hatchlings) of each species must be protected. Areas of particular concern include beaches between Melbourne Beach and Wabasso Beach, Florida, for loggerheads and the Rancho Nuevo beach in Mexico for Kemp's ridleys. Protection of nesting areas, nesting activities, and eggs and hatchlings is critical to the survival of the Kemp's ridley, and its importance for other species is increasing, in light of continued beach development, land use patterns, and other beach practices. Possible actions include public purchase of undeveloped beaches for restricted, nonthreatening uses; public purchase of development rights for undeveloped beaches; prohibition of vehicular traffic on beaches during nesting and incubation periods; control of lighting in the vicinity of nesting beaches; predator control; and establishment of a marine park at Rancho Nuevo. The 16 km of undeveloped beach property between Melbourne Beach and Wabasso Beach, Florida, in the Archie Carr National Wildlife Refuge proposed by the U.S. Fish and Wildlife Service, should be protected. Purchase of the land is the best method to ensure protection. The lands are available, and action should be taken before they are developed.

Small-scale, research-based captive breeding programs for Kemp's ridleys should be continued. Refinement of the technique would ensure the maintenance of a gene pool in captive animals in the event of a population loss. However, this option is not considered a promising management tool for the restoration of wild populations, because small populations of captive animals lack much of the genetic variability that was available in the wild population. In addition, the development of various survival behaviors, for example, feeding, nesting, and migration, might be impaired.

Headstarting should be maintained as a research tool, but cannot substitute for other essential conservation measures. The headstarting experiment should be continued, because it has research and public-awareness value. Experimental methods must continue to be improved. However, present knowledge makes it clear that headstarting, even if it works, will not be effective without simultaneous implementation of other conservation measures to reduce human-related deaths of juvenile and adult sea turtles.

Research

Sea Turtle Biology

In the process of evaluating the status of sea turtles, some knowledge gaps became apparent. Important data are missing—some difficult (or impossible) to obtain, some less so—that are imperative to good management. Current knowledge of sea turtles has allowed us to evaluate and recommend some basic research and conservation measures in this report so that further protection and recovery of sea turtle populations can be implemented.

Demographic Models For no species of sea turtle is knowledge of age-specific survivorship and age-specific fecundity adequate. Enough is known, however, about loggerhead demography to provide a fundamental understanding of basic concepts, such as the relative reproductive value of various life history stages. To evaluate fully the comparative importance of different sources of mortality and to evaluate the effectiveness of conservation measures, better information is needed on age at reproductive maturity, age-specific survivorship, age-specific fecundity, and their variances. Therefore, the committee recommends research on:

- Age-specific fecundity and survivorship, through enlargement of existing tagging programs and creation of new ones; special attention must be given to the tag-loss problem.
- Life stages and sex ratios, through increased efforts to count sea turtles of all age groups in as many habitats as possible; mortality estimates for all life stages are important.

Sea Turtle Distribution Information is needed on foraging habitats of sea turtles in deep water and the use of shallow water by juveniles and subadults. Sampling areas and times should be chosen to permit replication with the best current technology. Such surveys should help to define the amount of overlap of national jurisdictions and assist the implementation of cooperative programs, such as that between the United States and Mexico for the Kemp's ridley.

Sea Turtle Physiology and Pathology More information is needed on the effects on turtles of ingesting plastics, of petroleum products, of forced prolonged immersion, of cold-water stunning, of underwater explosions, and of oceanic debris. The committee recommends selected, complete necropsies to determine pathological conditions, causes of wounds, and any other cause of death. Research should also ascertain turtles' abilities to ingest plastics of various types and sizes without adverse effect. The effects of floating petroleum products on the reproductive and feeding behavior of turtles should be studied.

To allow for the complete evaluation of the use of tow-time limits in trawl fisheries, research should address how enforced submergence affects sea turtle anatomy and physiology as a function of season, water temperature, species and size, time of day, and history of previous forced submergence. Improved resuscitation techniques of comatose turtles should be developed from such research.

The effect of explosives on sea turtles during construction and demolition of marine structures or for any other reason (such as military ordnance) is largely unknown and must be investigated. Research should focus on the distribution and abundance of sea turtles near platforms designated for removal by explosives, confirmation and necropsy of dead turtles near explosion sites, the feasibility of moving turtles to different sites, and behavior of animals at explosion sites.

Research on the reproductive biology of sea turtles in the wild should continue.

Management Techniques

More research and experimentation are needed to improve TEDs and explore new alternatives. The techniques of deploying TEDs in a variety of conditions also need improvement. For example, it is important to reduce the tearing of trawl nets that have TEDs. If TEDs can be modified to allow efficient fishing for shrimp when seagrass and other algal detritus or other debris are abundant, a major objection to the use of TEDs by shrimp fishermen could be addressed. All these management techniques and options should include input from shrimp fishermen and gear experts. Enhancing acceptance of regulations on the shrimp fishing industry would enhance compliance and promote sea turtle conservation, and research should focus on whether education on TED use would be helpful in this regard. Research should also address other inducements to increase compliance with TED regulations.

There are strong grounds for believing that the drowning of sea turtles in trawls can be greatly reduced by the adoption of certain controls on the shrimp fishery, but it is important to evaluate the effectiveness of any regulations that are adopted. Because the shrimp fisheries are regulated to open and close at various specified dates at many places, the committee was able to use the timing of the fishery to test the impact of shrimp trawling on the numbers of stranded sea turtles.

This test produced strong evidence that shrimp trawling at some places and times is responsible for 70-80% of the sea turtles found stranded on the beaches of Texas and South Carolina. In contrast, the committee was not able to develop similar tests of the degree to which plastic debris, oil pollution, and other factors affect the survival of sea turtles.

The characteristics of the shrimp fishery that helped identify its effect on turtles should be used to test the effectiveness of the regulations. Historical data on the relationship of the numbers of stranded sea turtles to the opening and closing of the shrimp fishery in Texas and South Carolina should be collected in future years to evaluate the degree to which the drowning of sea turtles in trawls is reduced by the regulations.

Further research is necessary to assess the effectiveness of tow-time limitations. For example, are shrimp trawlers ever so concentrated that they repeatedly catch individual sea turtles often enough to make tow-time limits ineffective? Even though sea turtles can survive enforced submergence for some time, repeated submergence can cause drowning. Shrimp trawlers could help to answer the question by using carapace marks to denote captures and recaptures and then assessing turtle survival as a function of capture frequency during relatively short periods (e.g., a day). The results of the physiological research described above would also help.

The impact of fishing practices other than shrimp trawling on sea turtles might be large, but it is not well known. Research is needed on the impact of groundfish trawling, set-net and long-line fishing, gill nets, and pound-net fishing on sea turtles at different times and places.

Research on the complex effects of artificial protection of early life stages of sea turtles is needed. Special efforts should be directed to reproductive biology of captive sea turtles and the effects of rearing them in closed culture. Young turtles just out of captivity might not be prepared to survive in the wild. Research on means of acclimation of nursery-reared sea turtles would be profitable. How long should sea turtles be reared in captivity to maximize their ultimate survival in the wild? Is it worth taking an egg from a beach to raise the turtle in the nursery? It is not known whether female sea turtles are imprinted to nest on the beach where they were released. That information is needed before the place of release of nursery-reared turtles is determined.

The cumulative effects of human activities on nesting beaches should be quantified relative to the total available nesting areas, because the loss of nesting beaches through development or alteration could extirpate local populations. More research is needed on how to control or alter artificial lighting along nesting beaches, to minimize interference with nesting and with the crawl to the sea by emerging hatchlings. The impacts of motor vehicles on beaches, erosion control measures, and the development of beachfront property needs to be evaluated more completely.

References

Ackerman, R.A. 1980. Physiological and ecological aspects of gas exchange by sea turtle eggs. Am. Zool. 20:575-83.

Amos, A.F. 1989. Trash, debris and human activities: Potential hazards at sea and obstacles to Kemp's ridley sea turtle nesting. P. 42 in Proceedings of the First International Symposium on Kemp's Ridley Sea Turtle Biology, Conservation and Management, held October 1-4, 1985 in Galveston, Texas, C.W. Caillouet, Jr. and A.M. Landry, Jr., eds. TAMU-SG-89-105. Galveston: Sea Grant College Program, Texas A&M University.

Anonymous. 1976. Incidental capture of sea turtles by shrimp fishermen in Florida. Preliminary report of the Florida West Coast Survey, University of Florida Marine Advisory Program. 3 pp.

Anonymous. 1977. Alabama shrimp fishermen interviews for 1977-1978. Marine Resources Office, Alabama Cooperative Extension Service. 1 p.

Applied Biology, Inc. 1989a. Florida Power and Light Company, St. Lucie Unit 2: Annual Environmental Operating Report 1988. Atlanta, Ga.: Applied Biology, Inc. 82 pp.

Applied Biology, Inc. 1989b. Annual Activity Report for 1989 to Florida Department of Natural Resources. Atlanta, Ga.: Applied Biology, Inc. (Unpublished).

Balazs, G.H. 1980. Synopsis of Biological Data on the Green Turtle in the Hawaiian Islands. NOAA-TM-NMFS-SWFC-7. National Marine Fisheries, National Oceanic and Atmospheric Administration. 141 pp.

Balazs, G.H. 1982. Growth rates of immature green turtles in the Hawaiian Archipelago. Pp. 117-125 in Biology and Conservation of Sea Turtles, K.A. Bjorndal, ed. Washington, D.C.: Smithsonian Institution Press.

Balazs, G.H. 1983. Recovery Records of Adult Green Turtles Observed or Originally Tagged at French Frigate Shoals, Northwestern Hawaiian Islands. NOAA-TM-NMFS-SWFC-36. 42 pp.

References

Balazs, G.H. 1985. Impact of ocean debris on marine turtles: Entanglement and ingestion. Pp. 387-429 in Proceedings of the Workshop on the Fate and Impact of Marine Debris, 27-29 November, 1984, Honolulu, Hawaii, R.S. Shomura and H.O. Yoshida, eds. NOAA Tech. Memo. NMFS-NOAA-TM-NMFS-SWFC-54.

Balazs, G.H. 1986. Fibropapillomas in Hawaiian green turtles. Mar. Turtle Newsl. 39:1-3.

Baldwin, W.P., Jr., and J.P. Lofton. 1959. The loggerhead turtles of Cape Romain, South Carolina (abridged and annotated by D.K. Caldwell). Bull. Fla. State Mus. 4:319-348.

Barnard, D.E., J.A. Keinath, and J.A. Musick. 1989. Marine turtles (*Caretta caretta, Chelonia mydas, Dermochelys coriacea, and Lepidochelys kempi*) in Virginia and Adjacent Waters. Poster at First World Congress of Herpetology, held September 1989, at the University of Kent, Canterbury, England.

Bell, R., and J.I. Richardson. 1978. An analysis of tag recoveries from loggerhead sea turtles (*Caretta caretta*) nesting on Little Cumberland Island, Georgia. Pp. 20-24 in Proceedings of Florida and Interregional Conference on Sea Turtles, 24-25 July 1976, Jensen Beach, Florida. Fla. Mar. Res. Publ. No. 33, 66 pp.

Bellmund, S.A., P. Defur, C.W. Su, and J.A. Musick. 1985. Aromatic hydrocarbon analysis of Virginia sea turtles: Methods and implications. P. 56 in Proceedings of the Fifth Annual Workshop on Sea Turtle Biology and Conservation, J.I. Richardson, comp. Athens: Institute of Ecology, University of Georgia. (Unpublished).

Bellmund, S.A., J.A. Musick, R.C. Klinger, R.A. Byles, J.A. Keinath, and D.E. Barnard. 1987. Ecology of sea turtles in Virginia. Virginia Institute of Marine Science Special Scientific Report No. 119. Gloucester Pt., Va.: Virginia Institute of Marine Science. 49 pp.

Bjorndal, K.A., ed. 1981. Biology and Conservation of Sea Turtles. Proceedings of the World Conference on Sea Turtle Conservation, 26-30 November, 1979. Washington, D.C.: Smithsonian Institution Press.

Bjorndal, K.A. 1985. Nutritional ecology of sea turtles. Copeia 1985 (3):736-751.

Bjorndal, K.A., and A.B. Bolten. 1988. Growth rates of immature green turtles, *Chelonia mydas,* on feeding grounds in the southern Bahamas Atlantic Ocean. Copeia 1988 (3):555-564.

Bjorndal, K.A., and A. Carr. 1989. Variation in clutch size and egg size in the green turtle nesting population at Tortuguero, Costa Rica. Herpetologica 45:181-189.

Bjorndal, K.A., A.B. Meylan, and B.J. Turner. 1983. Sea turtles nesting at Melbourne Beach, Florida, I. Size, growth and reproductive biology. Biol. Conserv. 26:65-77.

Bjorndal, K.A., A. Carr, A.B. Meylan, and J.A. Mortimer. 1985. Reproductive biology of the hawksbill, *Eretmochelys imbricata,* at Tortuguero, Costa Rica, with notes on the ecology of the species in the Caribbean. Biol. Conserv. 34:353-368.

Boulon, R.H., Jr. 1983. The National Report for the Country of U.S. Virgin Islands. Pp. 489-499 in Proceedings of the Second Western Atlantic Turtle Symposium, L. Ogren et al., eds. NOAA Tech. Memo. NMFS-SEFC-226.

Brongersma, L.D. 1972. European Atlantic Turtles. Iitgegeven door het Rijksmuseum van Natuurlicke Historie te Leiden 121:1-318.

Brongersma, L.D. 1982. Marine turtles of the eastern Atlantic Ocean. Pp. 407-416 in Biology and Conservation of Sea Turtles, K.A. Bjorndal, ed. Washington, D.C.: Smithsonian Institution Press.

Bryant, T.L., ed. 1987. Stop plastic pollution. Delaware Sea Grant Reporter 6(3):6.

Bullis, H.R., Jr., and S.B. Drummond. 1978. Sea turtle captures off the southeastern United States by exploratory fishing vessels 1950-1976. Proceedings of the Florida and Interregional Conference on Sea Turtles, 24-25 July, Jensen Beach, Florida. Fla. Mar. Res. Pub. 33:45-50.

Bustard, H.R., and K.P. Tognetti. 1969. Green sea turtles: A discrete simulation of density-dependent population regulation. Science 163:939-941.

Byles, R.A. 1988. Behavior and Ecology of Sea Turtles from Chesapeake Bay, Virgina. Ph.D. dissertation. College of William and Mary, Williamsburg, Virginia. 121 pp.

Byles, R.A. 1989. Satellite telemetry of Kemp's ridley sea turtle, *Lepidochelys kempi,* in the Gulf of Mexico. Pp. 25-26 in Proceedings of the Ninth Annual Workshop on Sea Turtle Conservation and Biology held 7-11 February, 1989 at Jekyll Island, Georgia, S.A. Eckert, K.L.Eckert, and T.H. Richardson, comp. NOAA-TM-NMFS-SEFC-232. Miami, Fla.: Southeast Fisheries Center, National Marine Fisheries Service.

Calder, D.R., P.J. Eldridge, and E.B. Joseph. 1974. The Shrimp Fishery of the Southeastern United States: A Management Planning Profile. South Carolina Marine Resources Center, Technical Report No. 5, September, 1974.

Caldwell, D.K. 1959. The loggerhead turtles of Cape Romain, South Carolina. (Abridged and annotated manuscript of W.P. Baldwin, Jr. and J.P. Loftin, Jr.). Bull. Fla. State Mus. Biol. Sci. 4:319-348.

Caldwell, D.K., F.H. Berry, A. Carr, and R.A. Ragotzkie. 1959a. Multiple and group nesting by the Atlantic loggerhead turtle. Bull. Fla. State Mus. 4:309-318.

Caldwell, D.K., A. Carr, and L.H. Ogren. 1959b. Nesting and migration of the Atlantic loggerhead turtle. Bull. Fla. State Mus. 4:295-308.

Carr, A.F., Jr. 1952. Handbook of Turtles. The Turtles of the United States, Canada, and Baja California. Ithaca, N.Y.: Cornell University Press.

Carr, A.F., Jr. 1963. Panspecific reproductive convergence in *Lepidochelys Kempi*. Ergeb. Biol. 26:298-303.

Carr, A.F., Jr. 1967. So Excellent a Fishe. Garden City, N.Y.: Natural History Press. 249 pp.

Carr, A.F., Jr. 1973. The Everglades. The American Wilderness series. New York: Time-Life Books. 184 pp.

Carr, A.F., Jr. 1975. The Ascension Island green turtle colony. Copeia 1975 (3):547-555.

Carr, A.F., Jr. 1980. Some problems of sea turtle ecology. Am. Zool. 20:489-498.

Carr, A.F., Jr. 1986a. New Perspectives on the Pelagic Stage of Sea Turtle Development. NOAA-TM-SEFC-190. Miami, Fla.: Southeast Fisheries Center, National Marine Fisheries Service. 36 pp.

Carr, A.F., Jr. 1986b. Rips, FADS, and little loggerheads. BioScience 36:92-100.

Carr, A.F., Jr. 1987. The impact of nondegradable marine debris on the ecology and survival outlook of sea turtles. Mar. Pollut. Bull. 18(6B):352-356.

Carr, A.F., Jr., and A.B. Meylan. 1980. Evidence of passive migration of green turtle hatchlings in sargassum. Copeia 1980:366-368.

Carr, A.F., Jr., and L. Ogren. 1960. The ecology and migrations of sea turtles, IV: The green turtle in the Caribbean Sea. Bull. Am. Mus. Nat. Hist. 121:1-48.

Carr, A.F., Jr., and S. Stancyk. 1975. Observations on the ecology and survival outlook of the hawksbill turtle. Biol. Conserv. 8:161-172.

Carr, A.F., Jr., M.H. Carr, and A.B. Meylan. 1978. The ecology and migrations of sea turtles. Part 7. The West Caribbean green turtle colony. Bull. Am. Mus. Nat. Hist. 162:1-46.

Carr, A.F., Jr., L. Ogren, and C. McVea. 1981. Apparent hibernation by the Atlantic loggerhead turtle *Caretta caretta* off Cape Canaveral, Florida. Biol. Conserv. 19:7-14.

Chávez, H., M. Contreras, and T.P.E. Hernandez. 1967. Apectos biologicos y proteccion de la Tortuga Lora, *Lepidochelys kempi* (Garman), en la Costa de Tamaulipas, Mexico. Instituto Nacional Investigaciones Biologico-Pesqueras, Mexico 17:1-39.

Clark, D.R., Jr., and A.J. Krynitsky. 1980. Organo chlorine residues in eggs of loggerhead turtles (*Carerra caretta*) and green sea turtles (*Chelonia mydas*) nesting at Merritt Island, Florida, USA-July and August 1976. Pestic. Monit. J. 14:7-10.

Clark, D.R., Jr., and A.J. Krynitsky. 1985. DDE residues and artificial incubation of loggerhead sea turtle eggs. Bull. Environ. Contam. Toxicol. 34:121-125.

References

Clement Associates, Inc. 1989. Shrimp Trawling Requirements Rulemaking: Initial Technical Review. Fairfax, Va.: Clement Associates, Inc.

Cole, L.C. 1954. The population consequences of life history phenomena. Q. Rev. Biol. 29:103-137.

Coleman, E., ed. 1987. Marine litter: More than an eyesore. Aquanotes. Louisiana Sea Grant College Program 16(2):1-4.

Collard, S.B. 1987. Review of Oceanographic Features Relating to Neonate Sea Turtle Distribution and Dispersal in the Pelagic Environment: Kemp's Ridley (*Lepidochelys kempi*) in the Gulf of Mexico. Final report, NOAA-NMFS contract no. 40-GFNF-5-00193. National Marine Fisheries Service. 70 pp.

Conley, W.J., and B.A. Hoffman. 1987. Nesting activity of sea turtles in Florida, 1979-1985. Fla. Sci. 50:201-210.

Conner, D.K., 1987. Turtles, trawlers, and TEDs: What happens when the Endangered Species Act conflicts with fishermen's interests. Water Log (Coastal and Marine Law Research Program, University of Mississippi) 7:3-27.

Corliss, L.A., J.I. Richardson, C. Ryder, and R. Bell. 1989. The hawksbills of Jumby Bay, Antigua, West Indies. Pp. 33-35 in Proceedings of the Ninth Annual Workshop on Sea Turtle Conservation and Biology held 7-11 February, 1989 at Jekyll Island, Georgia, S.A. Eckert, K.L.Eckert, and T.H. Richardson, comp. NOAA-TM-NMFS-SEFC-232. Miami, Fla.: Southeast Fisheries Center, National Marine Fisheries Service.

Cornelius, S.E. 1986. The Sea Turtles of Santa Rosa National Park. San José, Costa Rica: Fundacion de Parques Nacionales. 64 pp.

Cottingham, D. 1988. Persistent marine debris: Challenge and response. The federal perspective. Alaska Sea Grant College Program. 41 pp.

Cox, B.A., and R.G. Mauermann. 1976. Incidential Catch and Disposal of Sea Turtles by the Brownsville-Port Isabel Gulf Shrimp Fleet. Texas Shrimp Association. (Unpublished).

Crouse, D.T., L.B. Crowder, and H. Caswell. 1987. A stage-based population model for loggerhead sea turtles and implications for conservation. Ecology 68:1412-1423.

Daniel, R.S., and K.U. Smith. 1947. Migration of newly hatched loggerhead turtles towards the sea. Science 106:398-399.

Danton, C., and R. Prescott. 1988. Kemp's ridley in Cape Cod Bay, Massachusetts—1987. Pp. 17-18 in Proceeding of the Eighth Annual Workshop on Sea Turtle Conservation and Biology held 24-26 February, 1988 at Fort Fisher, North Carolina, B.A. Schroeder, comp. NOAA-TM-NMFS-SEFC-214. Miami, Fla.: Southeast Fisheries Center, National Marine Fisheries Service.

Davis, G.E., and M.C. Whiting. 1977. Loggerhead sea turtle nesting in Everglades National Park, Florida, U.S.A. Herpetologica 33:18-28.

Dickerson, D.D., and D.A. Nelson. 1988. Use of long wavelength lights to prevent disorientation of hatchling sea turtles. Pp. 19-21 in Proceeding of the Eighth Annual Workshop on Sea Turtle Conservation and Biology held 24-26 February, 1988 at Fort Fisher, North Carolina, B.A. Schroeder, comp. NOAA-TM-NMFS-SEFC-214. Miami, Fla.: Southeast Fisheries Center, National Marine Fisheries Service.

Dickerson, D.D., and D.A. Nelson. 1989. Recent results on hatchling orientation responses in light wavelengths and intensities. Pp. 41-43 in Proceedings of the Ninth Annual Workshop on Sea Turtle Conservation and Biology held 7-11 February, 1989 at Jekyll Island, Georgia, S.A. Eckert, K.L. Eckert, and T.H. Richardson, comp. NOAA-TM-NMFS-SEFC-232. Miami, Fla.: Southeast Fisheries Center, National Marine Fisheries Service.

Dickerson, D.D., and D.A. Nelson, comps. 1990. Proceedings of the National Workshop on Methods to Minimize Dredging Impacts on Sea Turtles, 11 and 12 May 1988, Jacksonville, Florida. Misc. Paper EL-90-5. Vicksburg, Miss.: U.S. Army Corps of Engineers Waterways Experiment Station. 89 pp.

Dobie, J.L., L.H. Ogren, and J.F. Fitzpatrick, Jr. 1961. Food notes and records of the Atlantic ridley turtle (*Lepidochelys kempi*) from Louisiana. Copeia 1961:109-110.

Dodd, C.K., Jr. 1982. Nesting of the green turtle, *Chelonia mydas* (L.), in Florida: Historic review and present trends. Brimleyana 7:39-54.

Dodd, C.K., Jr. 1988. Synopsis of the Biological Data on the Loggerhead Sea Turtle *Caretta caretta* (Linnaeus 1758). USFWS Biological-88(14). Gainesville, Fla.: National Ecology Research Center. 119 pp. Available from NTIS as PB89-109565.

Doughty, R.W. 1984. Sea turtles in Texas: A forgotten commerce. Southwest. Hist. Q. 88:43-70.

Eckert, K.L. 1987. Environmental unpredictability and leatherback sea turtle (*Dermochelys coriacea*) nest loss. Herpetologica 43:315-323.

Eckert, K.L., and S.A. Eckert. 1983. Tagging and nesting research of leatherback sea turtles (*Dermochelys coriacea*) on Sandy Point, St. Croix, U.S. Virgin Islands, 1983. Final report submitted to the U.S. Fish and Wildlife Service, Jacksonville, Florida. (Unpublished).

Eckert, K.L., and S.A. Eckert, eds. 1988. Death of a giant. Dept. of Zoology, Univ. of Georgia. Mar. Turtle Newsl. 43:2-3.

Eckert, K.L., S.A. Eckert, and D.W. Nellis. 1984. Tagging and nesting research of leatherback sea turtles (*Dermochelys coriacea*) on Sandy Point, St. Croix, U.S. Virgin Islands, 1984, with a discussion of management options for the population. Report submitted to Division of Fish and Wildlife, Contract PC-CCA-178-84. United States Virgin Islands, Frederiksted, St. Croix. (Unpublished).

Edison Electric Institute. 1987. EEI Power Directory. Edison Electric Institute.

Eggers, J.M. 1989. Incidental capture of sea turtles at Salem Generating Station, Delaware Bay, New Jersey. Pp. 221-223 in Proceedings of the Ninth Annual Workshop on Sea Turtle Conservation and Biology held 7-11 February, 1989 at Jekyll Island, Georgia, S.A. Eckert, K.L. Eckert, and T.H. Richardson, comp. NOAA-TM-NMFS-SEFC-232. Miami, Fla.: Southeast Fisheries Center, National Marine Fisheries Service.

Ehrenfeld, D.W. 1968. The role of vision in the sea-finding orientation of the green turtle (*Chelonia mydas*). II. Orientation mechanism and range of spectral sensitivity. Anim. Behav. 16:281-287.

Ehrenfeld, D.W., and A.F. Carr. 1967. The role of vision in the sea-finding orientation of the green turtle (*Chelonia mydas*). Anim. Behav. 15:25-36.

Ehrhart, L.M. 1979. Reproductive characteristics and management potential of the sea turtle rookery at Canaveral National Seashore, Florida. Pp. 397-399 in Proceedings of the First Conference on Scientific Research in the National Parks, New Orleans, Louisiana, November 9-12, 1976, R.M. Linn, ed. Transactions and Proceedings Series-National Park Service, No. 5. Washington, D.C.: National Park Service, U.S. Government Printing Office.

Ehrhart, L.M. 1982. A review of sea turtle reproduction. Pp. 29-38 in Biology and Conservation of Sea Turtles, K.A. Bjorndal, ed. Washington, D.C.: Smithsonian Institution Press.

Ehrhart, L.M. 1983. Marine turtles of the Indian River Lagoon system. Fla. Sci. 46:337-346.

Farfante, I.P. 1969. Western Atlantic shrimps of the genus Penaeus. Fish. Bull. 67:461-591.

Farrell, J.G. 1988. Plastic pollution in the marine environment: Boaters can help control a growing problem. Delaware Sea Grant MAS Note. June. 1 p.

Ferris, J.S. 1986. Nest success and the survival and movement of hatchlings of the loggerhead sea turtle (*Caretta caretta*) on Cape Lookout National Seashore. CPSU Tech. Rep. 19. 40 pp.

FFOP (Northeast Region Foreign Fishery Observer Program). 1988. Summary of Japanese longline fishery, NW Atlantic Fishery Conservation Zone, calendar year 1987. Presented to Advisory Committee to U.S. Section of ICCAT, 26-27 October, Washington, D.C. 3 pp.

FFOP (Northeast Region Foreign Fishery Observer Program). 1989. Summary of Japanese longline fishery, NW Atlantic Fishery Conservation Zone, calendar year 1988. Presented to Advisory Committee to U.S. Section of ICCAT, 5-6 October, Washington, D.C. 3 pp.

Fisher, R.A. 1958. The Genetical Theory of Natural Selection, Second edition. New York: Dover.

Fletemeyer, J. 1980. A preliminary analysis of sea turtle eggs for DDE. Mar. Turtle Newsl. 15:6-7.

Fowler, L.E. 1979. Hatching success and nest predation in the green sea turtle, *Chelonia mydas,* at Tortuguero, Costa Rica. Ecology 60:946-955.

Frazer, N.B. 1983a. Survivorship of adult female loggerhead sea turtles, *Caretta caretta,* nesting on Little Cumberland Island, Georgia, USA. Herpetologica 39:436-447.

Frazer, N.B. 1983b. Demography and Life History Evolution of the Atlantic Loggerhead Sea Turtle, Caretta caretta. Ph.D. dissertation. University of Georgia, Athens, Georgia.

Frazer, N.B. 1984. A model for assessing mean age-specific fecundity in sea turtle populations. Herpetologica 40:281-291.

Frazer, N.B. 1986. Survival from egg to adulthood in a declining population of loggerhead turtles, *Caretta caretta.* Herpetologica 42:47-55.

Frazer, N.B. 1987. Preliminary estimates of survivorship for wild loggerhead sea turtles (*Caretta caretta*). J. Herpetol. 21:232-235.

Frazer, N.B., and L.M. Ehrhart. 1985. Preliminary growth models for green, *Chelonia mydas,* and loggerhead, *Caretta caretta,* turtles in the wild. Copeia 1985:73-79.

Frazer, N.B., and J.I. Richardson. 1985a. Seasonal variation in clutch size for loggerhead sea turtles, *Caretta caretta,* nesting on Little Cumberland Island, Georgia, USA. Copeia 1985:1083-1085.

Frazer, N.B., and J.I. Richardson. 1985b. Annual variation in clutch size and frequency for loggerhead turtles, *Caretta caretta,* nesting at Little Cumberland Island, Georgia, USA. Herpetologica 41:246-251.

Frazer, N.B., and J.I. Richardson. 1986. The relationship of clutch size and frequency to body size in loggerhead sea turtles, *Caretta caretta.* J. Herpetol. 20:81-84.

Frazier, J. 1984. Las tortugas marinas en el Oceano Atlantico Sur Occidental. Asoc. Herpetol. Argentina 2:2-21.

Frazier, J., and S. Salas. 1982. Tortugas marinas en Chile. Bol. Mus. Nac. Hist. Nat. Chile 39:63-73.

Fretey, J. 1982. Note sur les traumas observe chez des Tortues luths adultes *Dermochelys coriacea* (Vandelli) (Testudines, Dermochelyidae). Aquariol. 8(1981):119-128.

Fritts, T.H., and W. Hoffman. 1982. Diurnal nesting of marine turtles in southern Brevard County, Florida. J. Herpetol. 16:84-86.

Fritts, T.H., and M.A. McGehee. 1981. Effects of petroleum on the development and survival of marine turtle embryos. U.S. Fish and Wildlife Service, Denver Wildlife Research Center, Belle Chasse, La. FWS/OBS-81/37. 41 pp.

Fritts, T.H., and R.P. Reynolds. 1981. Pilot Study of the Marine Mammals, Birds and Turtles in OCS Areas of the Gulf of Mexico. FWS/OBS-81/36. Washington, D.C.: Fish and Wildlife Service, U.S. Department of the Interior.

Fritts, T.H., A.B. Irvine, R.D. Jennings, L.A. Collum, W. Hoffman, and M.A. McGehee. 1983. Turtles, Birds, and Mammals in the Northern Gulf of Mexico and Nearby Atlantic Waters. FWS/OBS-82/65. Washington, D.C.: Fish and Wildlife Service, U.S. Department of the Interior. 455 pp.

Gitschlag, G. 1989. Sea Turtle Monitoring at the Explosive Removal of Offshore Oil and Gas Structures: A Summary Report of the Sea Turtle Observer Program. Conducted by the National Marine Fisheries Service Galveston Laboratory, Galveston, Texas.

Glazebrook, J.S. 1980. Diseases of farmed sea turtles. Pp. 42-55 in Management of Turtle Resources, Applied Ecology Research Monograph 1. Queensland, Australia: James Cook University of North Queensland.

Graham, G. 1987. Final Report of the TED Project submitted to the Gulf and South Atlantic Fisheries Development Foundation, Inc., GASAFDFI No. 32-13-34500/20625. Texas A&M Marine Advisory Program, Texas A&M University System, College Station, Texas.

Gramentz, D. 1988. Involvement of loggerhead turtle with the plastic, metal, and hydrocarbon pollution in the Central Mediterranean. Mar. Poll. Bull. 19:11-13.

Grassman, M.A., and D. Owens. 1987. Chemosensory imprinting in juvenile green sea turtles (*Chelonia mydas*). Anim. Behav. 35:929-931.

Grassman, M.A., D.W. Owens, J.P. McVey, and R. Mrquez. 1984. Olfactory based orientation in artificially imprinted sea turtles. Science 224:83-84.

Gregg, S.S. 1988. Of soup and survival. Sea Front. 34:297-302.

Groombridge, B., comp. 1982. The IUCN Amphibia-Reptilia Red Data Book. Part I: Testudines, Crocodylia, Rhynchocephala. Gland, Switzerland: Species Survival Commission, International Union for Conservation of Nature and Natural Resources. 426 pp.

Gulf of Mexico Fishery Management Council. 1981. Fishery Management Plan for the Shrimp Fishery of the Gulf of Mexico, United States Waters. Gulf of Mexico Fishery Management Council, Tampa, Florida.

Hendrickson, J.R. 1958. The green sea turtle, *Chelonia mydas* (Linn.), in Malaya and Sarawak. Proc. Zool. Soc. Lond. 130:455-535.

Hendrickson, J.R. 1980. The ecological strategies of sea turtles. Am. Zool. 30:597-608.

Henwood, T.A. 1987. Movements and seasonal changes in loggerhead turtle *Caretta caretta* aggregations in the vicinity of Cape Canaveral, Florida (1978-84). Biol. Conserv. 40:191-202.

Henwood, T.A., and L.H. Ogren. 1987. Distribution and migrations of immature Kemp's ridley turtles (*Lepidochelys kempi*) and green turtles (*Chelonia mydas*) off Florida, Georgia and South Carolina. Northeast Gulf Sci. 9:153-159.

Henwood, T.A., and W.E. Stuntz. 1987. Analysis of sea turtle captures and mortalities during commercial shrimp trawling. Fish. Bull. 85:813-817.

Hildebrand, H.H. 1963. Hallazgo del area de anidacion de la tortuga morina "lora" *Lepidochelys kempi* (Garman) en la costa occidental del Golfo de Mexico. Ciencia 22:105-112.

Hildebrand, H.H. 1982. A historical review of the status of sea turtle populations in the western Gulf of Mexico. Pp. 447-453 in Biology and Conservation of Sea Turtles, K.A. Bjorndal, ed. Washington, D.C.: Smithsonian Institution Press.

Hillestad, H.O., R.J. Reimold, R.R. Stickney, H.L. Windom, and J.H. Jenkins. 1974. Pesticides, heavy metals and radionuclide uptake in loggerhead sea turtles from South Carolina and Georgia. Herpetol. Rev. 5:75.

Hillestad, H.O., J.I. Richardson, and G.K. Williamson. 1978. Incidental capture of sea turtles by shrimp trawlermen in Georgia. In Proceedings of the 32nd Annual Conference of the Southeastern Association of Fish and Wildlife Agencies. 5-8 November, Hot Springs, Virginia.

Hillis, Z-M., and A.L. Mackay. 1989a. Research Report on Nesting and Tagging of Hawksbill Sea Turtle (*Eretmochelys imbricata*) at Buck Island Reef National Monument. U.S. Virgin Islands, 1987-1988, Contract Report for Purchase Order No. PX5380-8-0090, U.S. National Park Service, Atlanta, Georgia. (Unpublished).

Hillis, Z-M., and A. L. Mackay. 1989b. Buck Island Reef National Monument sea turtle program, 1987-1988. Pp. 235-237 in Proceedings of the Ninth Annual Workshop on Sea Turtle Conservation and Biology held 7-11 February, 1989 at Jekyll Island, Georgia, S.A.

Eckert, K.L.Eckert, and T.H. Richardson, comp. NOAA-TM-NMFS-SEFC-232. Miami, Fla.: Southeast Fisheries Center, National Marine Fisheries Service.

Hirth, H.F. 1971. Synopsis of biological data on the green turtle *Chelonia mydas* (Linnaeus) 1758. FAO Fish. Synop. 85:1.1-8.19.

Hirth, H.F. 1980. Some aspects of the nesting behavior and reproductive biology of sea turtles. Am. Zool. 20:507-524.

Holland, D.F. 1989. Evaluation of Certified Trawl Efficiency Devices (TEDS) in North Carolina's Nearshore Ocean. Completion Report for Project 2-439-R. Raleigh: North Carolina Department of Natural Resources and Community Development.

Hopkins, S.R., and T.M. Murphy. 1983. Management of Loggerhead Turtle Nesting Beaches in South Carolina. Study Completion Report, E-1, Study No. VI-A-2, S.C. Wildlife and Marine Resources Department. 18 pp.

Hopkins, S., and J. Richardson. 1984. A recovery plan for marine turtles. Washington, D.C.: U.S. Government Printing Office.

Horikoshi, K. 1989. Egg survivorship of Tortuguero green turtles during the 1986 and 1988 seasons. Pp. 73-74 in Proceedings of the Ninth Annual Workshop on Sea Turtle Conservation and Biology held 7-11 February, 1989 at Jekyll Island, Georgia, S.A. Eckert, K.L. Eckert, and T.H. Richardson, comp. NOAA-TM-NMFS-SEFC-232. Miami, Fla.: Southeast Fisheries Center, National Marine Fisheries Service.

Hosier, P.E., M. Kochlar, and V. Thayer. 1981. Off-road vehicle and pedestrian track effects on the sea-approach of hatchling loggerhead turtles. Environ. Conserv. 8:158-161.

Huff, J.A. 1989. Florida (USA) terminates 'headstart' program. Mar. Turtle Newsl. 46:1-2.

Hughes, G.R. 1974. The sea turtles of South East Africa, 2. The biology of the Tongaland loggerhead turtle *Caretta caretta* L. with comments on the leatherback turtle *Dermochelys coriacea* L. and the green turtle *Chelonia mydas* L. in the study region. Oceanogr. Res. Inst. (Durban) Invest. Rep. 36:1-96.

Hughes, G.R. 1975. The marine turtles of Tongaland, VIII. Lammergeyer 22:9-18.

Jackson, D.R., E.E. Possardt, and L.M. Ehrhart. 1988. A joint effort to acquire critical sea turtle nesting habitat in east-central Florida. Pp. 39-41 in Proceeding of the Eighth Annual Workshop on Sea Turtle Conservation and Biology held 24-26 February, 1988 at Fort Fisher, North Carolina, B.A. Schroeder, comp. NOAA-TM-NMFS-SEFC-214. Miami, Fla.: Southeast Fisheries Center, National Marine Fisheries Service.

Jacobson, E.R., J.L. Mansell, J.P. Sundberg, L. Hajjar, M.E. Reichmann, L.M. Ehrhart, M. Walsh, and F. Murru. 1989. Cutaneous fibropapillomas of green turtles (*Chelonia mydas*). J. Comp. Pathol. 101:39-52.

Joyce, J.C. 1982. Protecting sea turtles. The Military Engineer, No. 481, July-August, p. 282-285.

Kaufmann, R. 1975. Studies on the loggerhead sea turtle, *Caretta caretta* (Linne) in Columbia, South America. Herpetologica 31:323-326.

Keinath, J.A., J.A. Musick, and R.A. Byles. 1987. Aspects of the biology of Virginia's sea turtles: 1979-1986. Va. J. Sci. 38:329-336.

Kemmerer, A.J. 1989. Summary Report from Trawl Tow Time Versus Sea Turtle Mortality Workshop. Southeast Fisheries Center, Pascagoula, Mississippi.

Killingly, J.S., and M. Lutcavage. 1983. Loggerhead turtle movements reconstructed from ^{18}O and ^{13}C profiles from commensal barnacle shells. Estuarine Coastal Shelf Sci. 16:345-349.

Kinne, O. 1985. Introduction to Volume IV, Part 2: Reptilia, aves, mammalia. Pp. 543-552 in Diseases of Marine Animals, Vol. IV, Part 2, Introduction, Reptilia, Aves, Mammalia, O. Kinne, ed. Hamburg, Federal Republic of Germany: Biologische Anstalt Helgoland.

Klima, E.F., and J.P. McVey. 1982. Headstarting the Kemp's ridley turtle, *Lepidochelys kempi*. Pp. 481-487 in Biology and Conservation of Sea Turtles, K.A. Bjorndal, ed. Washington, D.C.: Smithsonian Institution Press.

Klima, E.F., G.R. Gitschlag, and M.L. Renaud. 1988. Impacts of the explosive removal of offshore petroleum platforms on sea turtles and dolphins. Mar. Fish. Rev. 50:33-42.

Kontos, A.R. 1985. Sea Turtle Research Report, 1985: Mona Island, Puerto Rico. Report submitted to U.S. Fish and Wildlife Service, Unit Cooperative Agreement No. 14-16-009-1551, Jacksonville, Florida. (Unpublished).

Kontos, A.R. 1988. 1987 Annual Summary: Estimation of Sea Turtle Abundance: Mona Island, Puerto Rico. Report submitted to U.S. Fish and Wildlife Service, Unit Cooperative Agreement No. 14-16-009-1551, 12 May, 1988, Jacksonville, Florida. (Unpublished).

Kraemer, J.E. 1979. Variation in Incubation Length of Loggerhead Sea Turtle, *Caretta caretta,* Clutches on the Georgia Coast. M.S. thesis, University of Georgia, Athens.

Kraemer, J.E., and R. Bell. 1978. Rain-induced mortality of eggs and hatchlings of loggerhead sea turtles (*Caretta caretta*) on the Georgia Coast U.S.A. Herpetologica 36:72-77.

Laist, D.W. 1987. Overview of the biological effects of lost and discarded plastic debris in the marine environment. Mar. Poll. Bull. 18:319-326.

Lauckner, G. 1985. Diseases of reptilia. Pp. 543-552 in Diseases of Marine Animals, Vol. IV, Part 2, Introduction, Reptilia, Aves, Mammalia, O. Kinne, ed. Hamburg, Federal Republic of Germany: Biologische Anstalt Helgoland.

Lenarz, M.S., N.B. Frazer, M.S. Ralston, and R.B. Mast. 1981. Seven nests recorded for loggerhead turtle (*Caretta caretta*) in one season. Herpetol. Rev. 12:9.

Limpus, C.J. 1979. Notes on growth rates of wild turtles. Mar. Turtle Newsl. 10:3-5.

Limpus, C.J. 1985. A Study of the Loggerhead Sea Turtle, *Caretta caretta,* in Eastern Australia. Ph.D. dissertation. University of Queensland, Lucia, Australia. 481 pp.

Litwin, S.C. 1978. Loggerhead sea turtles of Jekyll Island, Georgia: A report on conservation. HERP Bull. N.Y. Herpetol. Soc. 14:18-21.

Lohoefener, R. 1988. Sea Turtles Offshore of the Chandeleur Islands, Louisiana: Apparent Preference for an Oil Platform Area. Southeast Fisheries Center, National Marine Fisheries Service, Pascagoula, Mississippi.

Lohoefener, R.R., W. Hoggard, C.L. Roden, K.D. Mullin, and C.M. Rogers. 1988. Distribution and relative abundance of surfaced sea turtles in the north-central Gulf of Mexico: Spring and fall 1987. Pp. 47-50 in Proceedings of the Eighth Annual Workshop on Sea Turtle Conservation and Biology held 24-26 February, 1988 at Fort Fisher, North Carolina, B.A. Schroeder, comp. NOAA-TM-NMFS-SEFC-214. Miami, Fla.: Southeast Fisheries Center, National Marine Fisheries Service.

Lohoefener, R.R., W. Hoggard, C.L. Roden, K.D. Mullin, and C.M. Rogers. 1989. Petroleum structures and the distribution of sea turtles. Pp. 31-35 in Proceedings of the Spring Ternary Gulf of Mexico Studies Meeting. U.S. Department of the Interior, Minerals Management Service, New Orleans, La.

Lotka, A.J. 1922. The stability of the normal age distribution. Proc. Natl. Acad. Sci. 8:339-345.

Lund, P.F. 1985. Hawksbill turtle (*Eretmochelys imbricata*) nesting on the east coast of Florida. J. Herp. 19:164-166.

Lutcavage, M. 1981. The Status of Marine Turtles in Chesapeake Bay and Virginia Coastal Waters. M.S. thesis. School of Marine Science, Virginia Institute of Marine Science, College of William and Mary. 127 pp.

Lutcavage, M., and J.A. Musick. 1985. Aspects of the biology of sea turtles in Virginia. Copeia 1985:449-456.

Lutz, P.L., and M. Lutcavage. 1989. The effects of petroleum on sea turtles: Applicability to Kemp's ridley. Pp. 52-54 in Proceedings of the First International Symposium on

Kemp's Ridley Sea Turtle Biology, Conservation and Management, held October 1-4, 1985 in Galveston, Texas, C.W. Caillouet, Jr. and A.M. Landry, Jr., eds. TAMU-SG-89-105. Galveston: Sea Grant College Program, Texas A&M University.

Mager, A., Jr. 1985. Five-year Status Reviews of Sea Turtles Listed Under the Endangered Species Act of 1973. St. Petersburg, Fla.: National Marine Fisheries Service. 90 pp.

Mann, T.M. 1977. Impact of developed coastline on nesting and hatchling sea turtles in southeastern Florida. M.S. thesis. Florida Atlantic University, Boca Raton, Florida. 100 pp.

Mann, T.M. 1978. Impact of developed coastline on nesting and hatchling sea turtles in southeastern Florida. Pp. 53-55 in Proceeding of the Florida and Interregional Conference on Sea Turtles, G.E. Henderson, ed. 24-25 July 1976, Jensen Beach, Florida. Fla. Mar. Res. Publ. No. 33. 66 pp.

Manzella, S.A., C.W. Caillouet, Jr., and C.T. Fontaine. 1988. Kemp's ridley, *Lepidochelys kempi,* sea turtle head start tag recoveries: Distribution, habitat, and method of recovery. Mar. Fish. Rev. 50:33-42.

Marcus, B.J., and C.G. Maley. 1987. Comparison of sand temperatures between a shaded and unshaded turtle nesting beach in South Florida. Seventh Annual Workshop on Sea Turtle Biology and Conservation. Wekiwa Springs State Park, Florida. (Unpublished).

Margaritoulis, D. 1982. Observations on loggerhead sea turtle *Caretta caretta* activity during three nesting seasons (1977-1979) in Zakynthos, Greece. Biol. Conserv. 24:193-204.

Márquez M., R. In prep. Synopsis of Biological Data on the Kemp's Ridley Sea Turtle *Lepidochelys kempi* (Garman, 1880). (Unpublished manuscript, in review).

Márquez M., R., A. Villaneuva O., and M. Sanchez P. 1981. The population of Kemp's ridley sea turtle in the Gulf of Mexico, *Lepidochelys kempi*. Pp. 159-164 in Biology and Conservation of Sea Turtles, K. Bjorndal, ed. Washington, D.C.: Smithsonian Institution Press.

Márquez M., R., M.S. Perez, J.D. Flores, and I. Arguello. 1989. Kemp's ridley research at Rancho Nuevo, 1987. Mar. Turtle Newsl. 44:6-7.

Mast, R.B., and J.L. Carr. 1985. Macrochelid mites in association with Kemp's ridley hatchlings. Mar. Turtle Newsl. 33:11-12.

McFarlane, R.W. 1963. Disorientation of loggerhead hatchlings by artificial road lighting. Copeia 1963:153.

McGavern, L. 1989. Plastics: Sealing the fact of marine life. MIT Sea Grant College Program for Coastweeks. 1 p.

McGehee, M.A. 1979. Factors Affecting the Hatching Success of Loggerhead Sea Turtle Eggs (*Caretta caretta*). M.S. thesis, University of Central Florida, Orlando.

McMurtray, J.D. 1982. Effect of screening sea turtle nests in a high predator density area. Report submitted to the School of Forest Resources, University of Georgia. (Unpublished).

McMurtray, J.D. 1986. Reduction of raccoon predation on sea turtle nests at Canaveral National Seashore, Florida. M.S. thesis. School of Forest Resources, University of Georgia. 48 pp.

McMurtray, J.D., and J.I. Richardson. 1985. A northern nesting record for the hawksbill turtle. Herp. Rev. 16:16-17.

Mendonca, M.T. 1981. Comparative growth rates of wild immature *Chelonia mydas* and *Caretta caretta* in Florida. J. Herpetol. 15:447-451.

Merrell, T.R., Jr. 1980. Accumulation of plastic litter on beaches of Amchitka Island, Alaska. Mar. Environ. Res. 3:171-184.

Mertz, D.B. 1970. Notes on methods used in life-history studies. Pp. 4-17 in Readings in Ecology and Ecological Genetics, J.H. Connell, D.B. Mertz, and W.W. Murdoch, eds. New York: Harper and Row.

Meylan, A.B. 1982. Sea turtle migrations: Evidence from tag returns. Pp. 91-100 in Biology and Conservation of Sea Turtles, K.A. Bjorndal, ed. Washington, D.C.: Smithsonian Institution Press.

Meylan, A.B. 1988. Spongivory in hawksbill turtles: A diet of glass. Science 239:393-395.

Meylan, A.B. 1989. Status report of the hawksbill turtle. Pp. 101-115 in Proceedings of the Second Western Turtle Symposium, L. Ogren et al., eds. NOAA Tech. Memo. NMFS-SEFC-226.

Meylan, A.B, and S. Sadove. 1986. Cold-stunning in Long Island Sound, New York. Mar. Turtle Newsl. 37:7-8.

Meylan, A.B., K.A. Bjorndal, and B.J. Turner. 1983. Sea turtles nesting at Melbourne Beach, Florida, II. Post-nesting movements of *Caretta caretta*. Biol. Conserv. 26:79-90.

MMS (Minerals Management Service). 1988. Federal Offshore Statistics: 1986. Leasing, Exploration, Production, & Revenues. OCS Report MMS 88-0010. U.S. Department of the Interior, Minerals Management Service, Vienna, Virginia.

Mo, C.L. 1988. Effect of Bacterial and Fungal Infection on Hatching Success of Olive Ridley Sea Turtle Eggs. Report to the U.S. World Wildlife Fund. 8 pp.

Morgan, P.J. 1989. Occurrence of leatherback turtles (*Dermochelys coriacea*) in the British Isles in 1988 with reference to a record specimen. Pp. 119-129 in Proceedings of the Ninth Annual Workshop of Sea Turtle Conservation and Biology, S.A. Eckert, K.L. Eckert, and T.H. Richardson, eds. 7-11 February, Jekyll Island, Georgia. NOAA Tech. Memo. NMFS-SEFC-232.

Morreale, S.J., and E.A. Standora. 1989. Occurrence, movement, and behavior of the Kemp's ridley and other sea turtles in New York waters. Okeanos Ocean Research Foundation Annual Report: April 1988 to April 1989. 35 pp.

Mortimer, J.A. 1982a. Feeding ecology of sea turtles. Pp. 103-109 in Biology and Conservation of Sea Turtles, K.A. Bjorndal, ed. Washington, D.C.: Smithsonian Institution Press.

Mortimer, J.A. 1982b. Factors influencing beach selection by nesting sea turtles. Pp. 45-51 in Biology and Conservation of the Sea Turtles, K.A. Bjorndal, ed. Washington, D.C.: Smithsonian Institution Press.

Mortimer, J.A. 1989. Research needed for management of the beach habitat. Pp. 236-246 in Proceedings of the Second Western Atlantic Turtle Symposium, L. Ogren et al., eds. NOAA Tech. Memo. NMFS-SEFC-226.

Mrosovsky, N. 1983. Ecology and nest-site selection of leatherback turtles, *Dermochelys coriacea*. Biol. Conserv. 26:47-56.

Mrosovsky, N., and S.F. Kingsmill. 1985. How turtles find the sea. Z. Tierpsychol. 67:237-256.

Murphy, T.M. 1985. Annual Performance Report. Submitted to the U.S. Fish and Wildlife Service, Jacksonville, Florida, by South Carolina Wildlife and Marine Resources Department, Columbia, South Carolina.

Murphy, T.M., and S.R. Hopkins. 1984. Aerial and Ground Surveys of Marine Turtle Nesting Beaches in the Southeast Region, U.S. Final report to National Marine Fisheries Service, Southeast Fisheries Center. 73 pp.

Murphy, T.M., and S.R. Hopkins-Murphy. 1989. Sea Turtle and Shrimp Fishing Interactions: A Summary and Critique of Relevant Information. Washington, D.C.: Center for Marine Conservation. 60 pp.

Musick, J.A. 1988. The sea turtles of Virginia with notes on identification and natural history. Second revised edition, Educational Series No. 24. Virginia Sea Grant College Program. 22 pp.

Nelson, D.A. 1986. Life history and environmental requirements of loggerhead sea turtles. U.S. Army Corps of Engineers Waterways Experiment Station, Vicksburg, Miss. Environ. Impact Res. Prog. Tech. Rep. EL-86-2.

Nelson, D.A., and D.D. Dickerson. 1989a. Effects of beach nourishment on sea turtles. Pp.

125-127 in Proceedings of the Ninth Annual Workshop on Sea Turtle Conservation and Biology held 7-11 February, 1989 at Jekyll Island, Georgia, S.A. Eckert, K.L. Eckert, and T.H. Richardson, comp. NOAA-TM-NMFS-SEFC-232. Miami, Fla.: Southeast Fisheries Center, National Marine Fisheries Service.

Nelson, D.A., and D.D. Dickerson. 1989b. Managment implications of recent hatchling orientation research. Pp. 129 in Proceedings of the Ninth Annual Workshop on Sea Turtle Conservation and Biology held 7-11 February, 1989 at Jekyll Island, Georgia, S.A. Eckert, K.L. Eckert, and T.H. Richardson, comp. NOAA-TM-NMFS-SEFC-232. Miami, Fla.: Southeast Fisheries Center, National Marine Fisheries Service.

Nelson, W.R., J. Benigno, and S. Burkett. 1987. Behavioral patterns of loggerhead sea turtles, *Caretta caretta,* in the Cape Canaveral area as determined by radio monitoring and acoustic tracking (abstract). Pp. 31 in Ecology of East Florida Sea Turtles: Proceedings of the Cape Canaveral, Florida, Sea Turtle Workshop, Miami, Florida, February 26-27, 1985. NOAA Tech. Rep. NMFS 53.

Nietschman, B. 1981. Following the underwater trail of a vanishing species: The Hawksbill turtle. Natl. Geogr. Soc. Res. Rep. 13:458-480.

NMFS (National Marine Fisheries Service). 1986a. Marine Recreational Fishery Statistics Survey, Atlantic and Gulf Coast, 1985: Current Fishery Statistics No. 8327. U.S. Department of the Interior, National Marine Fisheries Service.

NMFS (National Marine Fisheries Service). 1986b. Marine Recreational Fishery Statistics Survey, Pacific Coast, 1985: Current Fishery Statistics No. 8328. U.S. Department of the Interior, National Marine Fisheries Service.

NRC (National Research Council). 1987. Responding to Changes in Sea Level: Engineering Implications. Washington, D.C.: National Academy Press. 160 pp.

Ogren, L.H. 1989. Distribution of juvenile and subadult Kemp's ridley turtles: Preliminary results from the 1984-1987 surveys. Pp. 116-123 in Proceedings of the First International Symposium on Kemp's Ridley Sea Turtle Biology, Conservation and Management, held October 1-4, 1985 in Galveston, Texas, C.W. Caillouet, Jr. and A.M. Landry, Jr., eds. TAMU-SG-89-105. Galveston: Sea Grant College Program, Texas A&M University.

O'Hara, K.J., and S. Iudicello. 1987. Plastics in the Ocean: More than a Litter Problem. Center for Environmental Education. 128 pp.

O'Hara, K., N. Atkins, and S. Iudicello. 1986. Marine Wildlife Entanglement in North America. Prepared by Center for Environmental Education. 219 pp.

Olson, M.H. 1985. Population characteristics of the hawksbill turtle (*Eretmochelys imbricata*): A case study of the Endangered Species Act (Mona Island, Puerto Rico). Pp. 276-281 in Proceedings of the Fifth International Coral Reef Congress (Vol. 2), Tahiti.

Oravetz, C. 1989. Sea Turtle Conservation and Management Activities of the National Marine Fisheries Service, Southeast Region. 28 June. National Marine Fisheries Service. 26 pp.

Owens, D. 1980. The comparative reproductive physiology of sea turtles. Am. Zool. 20:546-563.

Owens, D. 1981. The role of reproductive physiology in the conservation of sea turtles. Pp. 39-44 in Biology and Conservation of Sea Turtles, K.A. Bjorndal, ed. Washington, D.C.: Smithsonian Institution Press.

Owens, D.W., and Y.A. Morris. 1985. The comparative endocrinology of sea turtles. Copeia 1985(3):723-735.

Owens, D.W., M.A. Grassman, and J.R. Hendrickson. 1982. The imprinting hypothesis and sea turtle reproduction. Herpetologica 38:124-135.

Parker, G.H. 1926. The growth of turtles. Proc. Natl. Acad. Sci. USA 12:422-424.

Parmenter, C.J. 1983. Reproductive migration in the hawksbill turtle (*Eretmochelys imbricata*). Copeia 1983:271-273.

Philibosian, R. 1976. Disorientation of hawksbill turtle hatchlings, *Eretmochelys imbricata,* by stadium lights. Copeia 1976:824.

Plotkin, P. 1989. Feeding ecology of the loggerhead sea turtle in the northwestern Gulf of Mexico. Pp. 139-141 in Proceedings of the Ninth Annual Workshop on Sea Turtle Conservation and Biology, S.A. Eckert, K.L. Eckert, and T.H. Richardson, eds. 7-11 February, Jekyll Island, Georgia. NOAA Tech. Memo. NMFS-SEFC-232.

Plotkin, P.T., and A.F. Amos. 1988. Entanglement in and ingestion of marine debris by sea turtles stranded along the south Texas Coast. Pp. 79-82 in Proceeding of the Eighth Annual Workshop on Sea Turtle Conservation and Biology held 24-26 February, 1988 at Fort Fisher, North Carolina, B.A. Schroeder, comp. NOAA-TM-NMFS-SEFC-214. Miami, Fla.: Southeast Fisheries Center, National Marine Fisheries Service.

Possardt, E.E., and D.R. Jackson. 1989. Status of proposed East-Central Florida sea turtle refuge. Pp. 143-144 in Proceedings of the Ninth Annual Workshop on Sea Turtle Conservation and Biology, S.A. Eckert, K.L. Eckert, and T.H. Richardson, eds. 7-11 February, Jekyll Island, Georgia. NOAA Tech. Memo. NMFS-SEFC-214.

Pritchard, P.C.H. 1971. The leatherback or leathery turtle, *Dermochelys coriacea.* IUCN Monogr. Mar. Turtle Ser. 1:1-39.

Pritchard, P.C.H. 1979. Encyclopedia of Turtles. Neptune, N.J.: T.F.H. Publishing. 895 pp.

Pritchard, P.C.H. 1980. The conservation of sea turtles: Practices and Problems. Am. Zool. 20:609-617.

Pritchard, P.C.H. 1990. Kemp's ridleys are rarer than we thought. Mar. Turtle Newsl. 49.

Pritchard, P.C.H., and R. Márquez M. 1973. Kemp's Ridley Turtle or Atlantic Ridley. International Union for the Conservation of Nature and Natural Resources Monograph No. 2 (Marine Turtle Series). 30 pp.

Pritchard, P.C.H., and T.H. Stubbs. 1982. An evaluation of sea turtle populations and survival status on Vieques Island. Final report submitted to Department of the Navy for Contract N6601-80-C-0560 (including modification No. P00001), Naval Ocean Systems Center, San Diego, Calif.

Pritchard, P.C.H., and P. Trebbau. 1984. The Turtles of Venezuela. Venezuela: Society for the Study of Amphibians and Reptiles. Contrib. Herpetol. No. 2. 403 pp.

Pruter, A.T. 1987. Sources, quantities and distribution of persistent plastics in the marine environment. Mar. Poll. Bull. 18:305-310.

Provancha, J.A., and L.M. Ehrhart. 1987. Sea turtle nesting trends at John F. Kennedy Space Center and Cape Canaveral Air Force Station, Florida, and relationships with factors influencing nest site selection. Pp. 33-44 in Ecology of East Florida Sea Turtles, W.N. Witzell, ed. NOAA Tech. Rep. NMFS 53. Miami, Fla.: Southeast Fisheries Center, National Marine Fisheries Service.

Rabalais, S.C., and N.N. Rabalais. 1980. The occurrence of sea turtles on the south Texas coast. Contrib. Mar. Sci. 23:123-129.

Ragotzkie, R.A. 1959. Mortality of loggerhead turtle eggs from excessive rainfall. Ecology 40:303-305.

Rayburn, R. 1986. Memorandum to respondents to turtle questionnaire. July 10, 1986. Texas Shrimp Association, Austin, Texas. 1 p.

Raymond, P.W. 1984. The Effects of Beach Restoration on Marine Turtles Nesting in South Brevard County, Florida. M.S. thesis. University of Central Florida, Orlando.

Richardson, J.I. 1990. The sea turtles of the King's Bay area and the endangered species observer program associated with construction dredging of the St. Mary's Entrance ship channel. Pp. 32-46 in Proceedings of the National Workshop on Methods to Minimize Dredging Impacts on Sea Turtles, D.A. Dickerson and D.A. Nelson, comp. 11-12 May 1988, Jacksonville, Florida. Miscellaneous Paper EL-90-5, U.S. Army Engineer Waterways Experiment Station, Vicksburg, Mississippi.

Richardson, J.I., and H.O. Hillestad. 1978. Ecology of a loggerhead sea turtle population. Pp. 22-37 in Proceedings, Rare and Endangered Wildlife Symposium, R.R. Odom and R. Landers, eds. Georgia Dept. Natural Resources, Tech. Bull. WL 4.

Richardson, J.I, and T.H. Richardson. 1982. An experimental population model for the loggerhead sea turtle (*Caretta caretta*). Pp. 165-174 in Biology and Conservation of Sea Turtles, K.A. Bjorndal, ed. Washington, D.C.: Smithsonian Institution Press.

Richardson, T.H., J.I. Richardson, C. Ruckdeschel, and M.W. Dix. 1978. Remigration patterns of loggerhead sea turtles (*Caretta caretta*) nesting on Little Cumberland and Cumberland Islands, Georgia. Fla. Mar. Res. Publ. 33:39-44.

Roithmayr, C., and T. Henwood. 1982. Incidental catch and mortality report. Final report to Southeast Fisheries Center. National Marine Fisheries Service, NOAA, Miami, Fla. 20 pp.

Rosman, I., G.S. Boland, L. Martin, and C. Chandler. 1987. Underwater Sightings of Sea Turtles in the Northern Gulf of Mexico. U.S. Department of the Interior, Minerals Management Service, Gulf of Mexico OCS Regional Office.

Ross, J.P. 1982. Historical decline of loggerhead, ridley and leatherback sea turtles. Pp. 189-195 in Biology and Conservation of Sea Turtles, K.A. Bjorndal, ed. Washington, D.C.: Smithsonian Instituion Press.

Ross, J.P., and M.A. Barwani. 1982. Review of sea turtles in the Arabian area. Pp. 373-383 in Biology and Conservation of Sea Turtles, K.A. Bjorndal, ed. Washington, D.C.: Smithsonian Institution Press.

Ross, J.P., S. Beavers, D. Mundell, and M. Airth-Kindree. 1989. The Status of Kemp's Ridley. A report to the Center for Marine Conservation from Caribbean Conservation Corporation. 51 pp.

Rostal, D.C., D.W. Owens, E.E. Louis, and M.S. Amoss. 1988. The reproductive biology of captive Kemp's ridleys (*Lepidochelys kempi*): Pilot studies. Pp. 91-96 in Proceedings of the Eighth Annual Workshop on Sea Turtle Conservation and Biology held 24-26 February, 1988 at Fort Fisher, North Carolina, B.A. Schroeder, comp. NOAA-TM-NMFS-SEFC-214. Miami, Fla.: Southeast Fisheries Center, National Marine Fisheries Service.

Rudloe, J. 1981. From the jaws of death. Sports Illus. 54(13):60-70.

Sadove, S.S., and S.J. Morreale. 1989. Marine mammal and sea turtle encounters with marine debris in the New York Bight and the Northeast Atlantic. Draft. Okeanos Ocean Research Foundation. 12 pp.

Salmon, M., and J. Wyneken. 1987. Orientation and swimming behavior of hatchling loggerhead sea turtles (*Caretta caretta*) during their off-shore migration. J. Exp. Mar. Biol. Ecol. 109:137-153.

Sanders, M. 1989. Marine debris: Fatal attraction. MAKAI. Univ. of Hawaii Sea Grant Coll. Progr. 11:2-3.

San Francisco Chronicle. January 8, 1983. Turtles eat plastic bags as jellyfish. p. 24.

Schmelz, G.W., and R.R. Mezich. 1988. A preliminary investigation of the potential impact of Australian pines on the nesting activities of the loggerhead turtle. Pp. 63-66 in Proceedings of the Eighth Annual Workshop on Sea Turtle Conservation and Biology held 24-26 February, 1988 at Fort Fisher, North Carolina, B.A. Schroeder, comp. NOAA-TM-NMFS-SEFC-214. Miami, Fla.: Southeast Fisheries Center, National Marine Fisheries Service.

Schroeder, B.A. 1987. 1986 Annual Report of the Sea Turtle Stranding and Salvage Network. Atlantic and Gulf Coasts of the United States January—December 1986. NOAA-NMFS-SEFC. CRD-87/88-12. Miami, Fla.: Coastal Resources Division, National Marine Fisheries Service. 45 pp.

Schroeder, B.A., and C.A. Maly. 1989. 1988 fall/winter strandings of marine turtles along the northeast Florida and Georgia coasts. Pp. 159-161 in Proceedings of the Ninth Annual Workshop on Sea Turtle Conservation and Biology held 7-11 February, 1989 at Jekyll

Island, Georgia, S.A. Eckert, K.L. Eckert, and T.H. Richardson, comp. NOAA-TM-NMFS-SEFC-232. Miami, Fla.: Southeast Fisheries Center, National Marine Fisheries Service.

Schroeder, B.A., and A.A. Warner. 1988. 1987 Annual Report of the Sea Turtle Stranding and Salvage Network. Atlantic and Gulf Coasts of the United States January - December 1987. NOAA-NMFS-SEFC. CRD-87/88-28. Miami, Fla.: Coastal Resources Division, National Marine Fisheries Service. 45 pp.

Schulz, J.P. 1975. Sea turtles nesting in Surinam. Zool. Verh. (Leiden) No. 143:3-143.

Shaver, D.J., D.W. Owens, A.H. Chaney, C.W. Caillouet, Jr., P. Burchfield, and R. Márquez M. 1988. Styrofoam box and beach temperatures in relation to incubation and sex ratios of Kemp's ridley sea turtles. Pp. 103-108 in Proceedings of the Eighth Annual Workshop on Sea Turtle Conservation and Biology held 24-26 February, 1988 at Fort Fisher, North Carolina, B.A. Schroeder, comp. NOAA-TM-NMFS-SEFC-214. Miami, Fla.: Southeast Fisheries Center, National Marine Fisheries Service.

Shoop, C.R., and C. Ruckdeschel. 1982. Increasing turtle strandings in the southeast United States: A complicating factor. Biol. Conserv. 23:213-215.

Slay, C.K., and J.I. Richardson. 1988. King's Bay, Georgia: Dredging and Turtles. Pp. 109-111 in Proceedings of the Eighth Annual Workshop on Sea Turtle Conservation and Biology held 24-26 February, 1988 at Fort Fisher, North Carolina, B.A. Schroeder, comp. NOAA-TM-NMFS-SEFC-214. Miami, Fla.: Southeast Fisheries Center, National Marine Fisheries Service.

Small, V. 1982. Sea Turtle Nesting at Virgin Islands National Park and Buck Island Reef National Monument, 1980 and 1981. R/RM/SER-61. St. Thomas: Virgin Islands National Park. 65 pp. Available from NTIS as PB83-137406.

Smith, G.M., and C.W. Coates. 1938. Fibro-epithelial growths of the skin in large marine turtles, *Chelonia mydas* (Linnaeus). Zoologica 23:93-98.

Squires, H.J. 1954. Records of marine turtles in the Newfoundland area. Copeia 1954:68.

Stancyk, S.E. 1982. Non-human predators of sea turtles and their control. Pp. 139-152 in Biology and Conservation of Sea Turtles, K.A. Bjorndal, ed. Washington, D.C.: Smithsonian Institution Press.

Standora, E.A., and J.R. Spotila. 1985. Temperature dependent sex determination in sea turtles. Copeia 1985 (3):711-722.

Standora, E.A., S.J. Morreale, R. Estes, R. Thompson, and M. Hilburger. 1989. Growth rates of juvenile Kemp's ridleys and their movement in New York waters. Pp. 175-177 in Proceedings of the Ninth Annual Workshop on Sea Turtle Conservation and Biology held 7-11 February, 1989 at Jekyll Island, Georgia, S.A. Eckert, K.L.Eckert, and T.H. Richardson, comp. NOAA-TM-NMFS-SEFC-232. Miami, Fla.: Southeast Fisheries Center, National Marine Fisheries Service.

Stanley, K.M., E.K. Stabenau, and A.M. Landry. 1988. Debris ingestion by sea turtles along the Texas coast. Pp. 119-121 in Proceedings of the Ninth Annual Workshop on Sea Turtle Conservation and Biology, S.A. Eckert, K.L. Eckert, and T.H. Richardson, eds. 7-11 February, Jekyll Island, Georgia. NOAA Tech. Memo. NMFS-SEFC-214.

Stoneburner, D.L., M.N. Nicora, and E.R. Blood. 1980. Heavy metals in loggerhead sea turtle eggs (*Caretta caretta*): Evidence to support the hypothesis that demes exist in the western Atlantic population. J. Herpetol. 14:171-176.

Street, M.W. 1987. Evaluation of Reported Sea Turtle Strandings and Shrimp Trawl Data for North Carolina. Report. No. 166. Raleigh: Division of Marine Fisheries, North Carolina Department of Natural Resources and Community Development.

Studt, J.F. 1987. Amelioration of maintenance dredging impacts on sea turtles, Canaveral Harbor, Florida. Pp. 55-58 Ecology of East Florida Sea Turtles: Proceedings of the Cape Canaveral, Florida, Sea Turtle Workshop, Miami, Florida, February 26-27, 1985. NOAA Tech. Rep. NMFS 53.

Stuntz, W.E., and A.J. Kemmerer. 1989. Comparison of Sea Turtle Mortality Under Tow Time and TED Managment Options: A Special Report. Mississippi Laboratories, Southeast Fisheries Center, National Marine Fisheries Service, Pascagoula, Miss.

Swordfish Management Plan. 1985. Fishery Management Plan, Regulatory Impact Review, Initial Regulatory Flexibility Analyses, and Final Environmental Impact Statement for Atlantic Swordfish. Prepared by the South Atlantic Fishery Management Council in Cooperation with Caribbean FMC, Gulf of Mexico FMC, Mid-Atlantic FMC, and New England FMC. February. Charleston, S.C. 306 pp.

Talbert, O.R., Jr., S.E. Stancyk, J.M. Dean, and J.M. Will. 1980. Nesting activity of the loggerhead turtle (*Caretta caretta*) in South Carolina. I: A rookery in transition. Copeia 1980:709-719.

Tambiah, C.R. 1989. Sea Turtles of Mona Island, Puerto Rico: Annual Research Report: 1988. Report submitted to U.S. Fish and Wildlife Service, Unit Cooperative Agreement No. 14-16-009-1551, Jacksonville, Florida, 1 April 1989. (Unpublished).

Teas, W.G., and A. Martinez. 1989. 1988 Annual Report of the Sea Turtle Stranding and Salvage Network Atlantic and Gulf Coasts of the United States January - December 1988. Contribution No. CRD-88/89-19. Miami, Fla.: Coastal Resources Division, National Marine Fisheries Service.

Thompson, N.B. 1984. Progress Report on Estimating Density and Abundance of Marine Turtles: Results of First Year Pelagic Surveys in the Southeast U.S. National Marine Fisheries Service, Miami, Fla., January 1984. 30 pp. and Figures and Tables.

Thompson, N.B. 1988. The status of loggerhead, *Caretta caretta*; Kemp's ridley, *Lepidochelys kempi*; and green, *Chelonia mydas*, sea turtles in U.S. waters. Mar. Fish. Rev. 50:16-23.

Thompson, N.P., P.W. Rankin, and D.W. Johnston. 1974. Polychlorinated biphenyls and *p,p*-DDE in green turtle eggs from Ascension Island, South Atlantic Ocean. Bull. Environ. Contam. Toxicol. 11:399-406.

Thurston, J., and T.A. Wiewandt. 1975. Management of sea turtles on Mona Island: Appendix I. Report to the Puerto Rico Department of Natural Rsources, San José, Costa Rica. 17 pp.

Timko, R.E., and L. Kolz. 1982. Satellite sea turtle tracking. Mar. Fish. Rev. 44:19-24.

Tucker, A.D. 1989a. So many turtles, so little time: Understanding fecundity and overestimating populations? Pp. 181-183 in Proceedings of the Ninth Annual Workshop on Sea Turtle Conservation and Biology held 7-11 February, 1989 at Jekyll Island, Georgia, S.A. Eckert, K.L. Eckert, and T.H. Richardson, comp. NOAA-TM-NMFS-SEFC-232. Miami, Fla.: Southeast Fisheries Center, National Marine Fisheries Service.

Tucker, A.D. 1989b. The influence of reproductive variation and spatial distribution on nesting success for leatherback sea turtles (*Dermochelys coriacea*). M.S. thesis. Department of Zoology, University of Georgia. 122 pp.

Ulrich, G.F. 1978. Incidential Catch of Loggerhead Turtles by South Carolina Commercial Fisheries. Report to NMFS, Contract No. 03-7-042-35151 and 03-7-042-35121. 48 pp.

Van Meter, V. 1983. Florida's Sea Turtles. Miami, Fla.: Florida Light and Power Co. 44 pp.

van Nierop, M.M., and J.C. den Hartog. 1984. A study of the gut contents of five juvenile loggerhead turtles, *Caretta caretta* (Linnaeus) (Reptilia, Cheloniidae), from the southeastern part of the north Atlantic Ocean, with emphasis on coelenterate identification. Zool. Meded. Leiden 59:35-54.

Wallace, N. 1985. Debris entanglement in the marine environment: A review. Pp. 259-277 in Proceedings of the Ninth Annual Workshop on Sea Turtle Conservation and Biology held 7-11 February, 1989 at Jekyll Island, Georgia, S.A. Eckert, K.L. Eckert, and T.H. Richardson, comp. NOAA-TM-NMFS-SEFC-232. Miami, Fla.: Southeast Fisheries Center, National Marine Fisheries Service.

Wehle, D.H.S., and F.C. Coleman. 1983. Plastics at sea. Nat. Hist. 92:20-26.

Welch, L., ed. 1988. Marine Log. Florida Sea Grant College Program. October, p. 3.

Werner, P.A., and H. Caswell. 1977. Population growth rates and age- versus stage-distribution models for teasel (*Dipsacus sylvestris* Huds.). Ecology 58:1103-1111.

Whitmore, C.P., and P.H. Dutton. 1985. Infertility, embryonic mortality and nest-site selection in leatherback (*Dermochelys-coriacea*) and green sea turtles (*Chelonia-mydas*) in Suriname. Biol. Conserv. 34:251-272.

Wibbels, T., N. Frazer, M. Grassman, J.R. Hendrickson, and P.C.H. Pritchard. 1989. Blue Ribbon Panel Review of the National Marine Fisheries Service Kemp's Ridley Headstart Program. Submitted to Southeast regional office, NMFS. August 1989. Miami, Fla.: Southeast Fisheries Center, National Marine Fisheries Service. 13 pp.

Wilcox, W.A. 1986. Commercial fisheries of the Indian River, Florida. Rept. U.S. Comm. Fish Fish. 22:249-262.

Winn, H.E. 1982. A Characterization of Marine Mammals and Turtles in the Mid-and-North Atlantic Areas of the U.S. Outer Continental Shelf. Final Report of the Cetacean and Turtle Assessment Program, under Contract No. AA551-CT8-48. Kingston: University of Rhode Island.

Witham, P.R. 1978. Does a problem exist relative to small sea turtles and oil spills? Pp. 629-632 in Proceedings of the Conference on Assessment of Ecological Impacts of Oil Spills, 14-17 June 1978, Keystone, Colo. Washington, D.C.: American Institute of Biological Sciences, Inc.

Witham, R., and C.R. Futch. 1977. Early growth and oceanic survival of pen-reared sea turtles. Herpetologica 33:404-409.

Witherington, B.E. 1986. Human and Natural Causes of Marine Turtle Clutch and Hatchling Mortality and Their Relationship to Hatchling Production on an Important Florida Nesting Beach. M.S. thesis. University of Central Florida, Orlando, Florida. 141 pp.

Witherington, B.E., and L.M. Ehrhart. 1989a. Status and reproductive characteristics of green turtles (*Chelonia mydas*) nesting in Florida. Pp. 351-352 in Proceedings of the Second Western Atlantic Turtle Symposium held 12-16 October, 1987 in Mayaguez, Puerto Rico, L. Ogren, ed. NOAA-TM-NMFS-SEFC-226. Panama City, Fla.: Panama City Laboratory, National Marine Fisheries Service. Available from NTIS as PB90-127648.

Witherington, B.E., and L.M. Ehrhart. 1989b. Hypothermic stunning and mortality of marine turtles in the Indian River Lagoon system, Florida, U.S.A. Copeia 1989 (3):696-703.

Witkowski, S.A., and J.G. Frazier. 1982. Heavy metals in sea turtles. Mar. Pollut. Bull. 13:254-255.

Witzell, W.N. 1983. Synopsis of biological data on the hawksbill turtle *Eretmochelys imbricata* (Linnaeus, 1766). FAO Fish. Synop. 137:78.

Witzell, W.N. 1987. Selective predation on large cheloniid sea turtles by tiger sharks (*Galeocerdo cuvier*). Jpn. J. Herpetol. 12:22-29.

Wolf, R.E. 1989. Boca Raton sea turtle protection program (1988) in conjunction with the North Beach Nourishment Project. Pp. 191-192 in Proceedings of the Ninth Annual Workshop on Sea Turtle Conservation and Biology held 7-11 February, 1989 at Jekyll Island, Georgia, S.A. Eckert, K.L. Eckert, and T.H. Richardson, comp. NOAA-TM-NMFS-SEFC-232. Miami, Fla.: Southeast Fisheries Center, National Marine Fisheries Service.

Wolke, R.E. 1989. Pathology and Sea Turtle Conservation. Report submitted to National Academy of Sciences/National Research Council Committee on Sea Turtle Conservation, Washington, D.C. 25 pp.

Wolke, R.E., D.R. Brooks, and A. George. 1982. Spirorchidiasis in loggerhead sea turtles (*Caretta caretta*): Pathology. J. Wildl. Dis. 18:175-186.

Wood, J.R., and F.E. Wood. 1980. Reproductive biology of captive green sea turtles *Chelonia mydas*. Am. Zool. 20:499-505.

Wood, J.R., and F.E. Wood. 1984. Captive breeding of the Kemp's ridley. Mar. Turtle Newsl. 29:12.

Wyneken, J., T.J. Burke, M. Salmon, and D.K. Pedersen. 1988. Egg failure in natural and relocated sea turtle nests. J. Herpetol. 22:88-96.

Yntema, C.I., and N. Mrosovsky. 1980. Sexual differentiation in hatchling loggerheads (*Caretta caretta*) incubated at different controlled temperatures. Herpetologica 36:33-36.

Yntema, C.I., and N. Mrosovsky. 1982. Critical periods and pivotal temperatures for sexual differentiation in loggerhead sea turtles. Can. J. Zool. 60:1012-1016.

Appendixes

Appendix A

Endangered Species Act Amendments of 1988

Public Law 100-478
100th Congress

An Act

Oct. 7, 1988
[H.R. 1467]

To authorize appropriations to carry out the Endangered Species Act of 1973 during fiscal years 1988, 1989, 1990, 1991, and 1992, and for other purposes.

Wildlife.
Conservation.

Be it enacted by the Senate and House of Representatives of the United States of America in Congress assembled,

TITLE I—ENDANGERED SPECIES ACT AMENDMENTS OF 1988

SEC. 1001. DEFINITIONS.

(a) DEFINITION OF PERSON.—Paragraph (13) of section 3 of the Endangered Species Act (16 U.S.C. 1532) is amended to read as follows:

"(13) The term 'person' means an individual, corporation, partnership, trust, association, or any other private entity; or any officer, employee, agent, department, or instrumentality of the Federal Government, of any State, municipality, or political subdivision of a State, or of any foreign government; any State, municipality, or political subdivision of a State; or any other entity subject to the jurisdiction of the United States.".

(b) DEFINITION OF SECRETARY.—Paragraph (15) of section 3 of the Endangered Species Act (16 U.S.C. 1532) is amended by inserting "also" before "means the Secretary of Agriculture".

SEC. 1008. SEA TURTLE CONSERVATION.

(a) DELAY OF REGULATIONS.—The Secretary of Commerce shall delay the effective date of regulations promulgated on June 29, 1987, relating to sea turtle conservation, until May 1, 1990, in inshore areas, and until May 1, 1989, in offshore areas, with the exception that regulations already in effect in the Canaveral area of Florida shall remain in effect. The regulations for the inshore area shall go into effect beginning May 1, 1990, unless the Secretary determines that other conservation measures are proving equally effective in reducing sea turtle mortality by shrimp trawling. If the Secretary makes such a determination, the Secretary shall modify the regulations accordingly.

(b) STUDY.—

(1) IN GENERAL.—The Secretary of Commerce shall contract for an independent review of scientific information pertaining to the conservation of each of the relevant species of sea turtles to be conducted by the National Academy of Sciences with such

individuals not employed by Federal or State government other than employees of State universities and having scientific expertise and special knowledge of sea turtles and activities that may affect adversely sea turtles.

(2) PURPOSES OF REVIEW.—The purposes of such independent review are—

(i) to further long-term conservation of each of the relevant species of sea turtles which occur in the waters of the United States;

(ii) to further knowledge of activities performed in the waters and on the shores of the United States, Mexico and other nations which adversely affect each of the relevant species of sea turtles;

(iii) to determine the relative impact which each of the activities found to be having an adverse effect on each of the relevant species of turtles has upon the status of each such species;

(iv) to assist in identifying appropriate conservation and recovery measures to address each of the activities which affect adversely each of the relevant species of sea turtles;

(v) to assist in identifying appropriate reproductive measures which will aid in the conservation of each of the relevant species of sea turtles;

(vi) in particular to assist in determining whether more or less stringent measures to reduce the drowning of sea turtles in shrimp nets are necessary and advisable to provide for the conservation of each of the relevant species of sea turtles and whether such measures should be applicable to inshore and offshore areas as well as to various geographical locations; and

(vii) to furnish information and other forms of assistance to the Secretary for his use in reviewing the status of each of the relevant species of sea turtles and in carrying out other responsibilities contained under this Act and law.

(3) SCOPE OF REVIEW.—The terms and outlines of such independent review shall be determined by a panel to be appointed by the President of the National Academy of Sciences, except that such review shall include, at a minimum, the following information:

(i) estimates of the status, size, age structure and, where possible, sex structure of each of the relevant species of sea turtles;

(ii) the distribution and concentration, in terms of United States geographic zones, of each of the relevant species of sea turtles;

(iii) the distribution and concentration of each of the relevant species of sea turtles, in the waters of the United States, Mexico and other nations during the developmental, migratory and reproductive phases of their lives;

(iv) identification of all causes of mortality, in the waters and on the shores of the United States, Mexico and other nations for each of the relevant species of sea turtles;

(v) estimates of the magnitude and significance of each of the identified causes of turtle mortality;

(vi) estimates of the magnitude and significance of present or needed head-start or other programs designed to increase the production and population size of each of the relevant species of sea turtles;

(vii) description of the measures taken by Mexico and other nations to conserve each of the relevant species of sea turtles in their waters and on their shores, along with a description of the efforts to enforce these measures and an assessment of the success of these measures;

(viii) the identification of nesting and/or reproductive locations for each of the relevant species of sea turtles in the waters and on the shores of the United States, Mexico and other nations and measures that should be undertaken at each location as well as a description of worldwide efforts to protect such species of turtles.

(4) COMPLETION AND SUBMISSION OF REVIEW.—Such independent review shall be completed after an opportunity is provided for individuals with scientific and special knowledge of sea turtles and activities that may affect adversely sea turtles to present relevant information to the panel. It shall then be submitted by the Secretary, together with recommendations by the Secretary in connection therewith, to the Committee on Environment and Public Works of the United States Senate and the Committee on Merchant Marine and Fisheries of the United States House of Representatives on or before April 1, 1989. In the event the independent review cannot be completed by April 1, 1989, then the panel shall give priority to completing the independent review as it applies to the Kemp's ridley sea turtle and submitting the same to the Secretary by that date, or as expeditiously as possible, and thereafter shall complete as expeditiously as possible the remaining work of the independent review.

(5) REVIEW OF STATUS.—After receipt of any portion of the independent review from the panel, the Secretary shall review the status of each of the relevant species of sea turtles.

(6) RECOMMENDATIONS OF SECRETARY.—The Secretary, after receipt of any portion of the independent review from the panel, shall consider, along with the requirements of existing law, the following before making recommendations:

(i) reports from the panel conducting the independent review;

(ii) written views and information of interested parties;

(iii) the review of the status of each of the relevant species of sea turtles;

(iv) the relationship of any more or less stringent measures to reduce the drowning of each of the relevant species of sea turtles in shrimp nets to the overall conservation plan for each such species;

(v) whether increased reproductive or other efforts in behalf of each of the relevant species of sea turtles would make no longer necessary and advisable present or proposed conservation regulations regarding shrimping nets;

(vi) whether certain geographical areas such as, but not

limited to, inshore areas and offshore areas, should have more stringent, less stringent or different measures imposed upon them in order to reduce the drowning of each of the relevant species of sea turtles in shrimp nets;

(vii) other reliable information regarding the relationship between each of the relevant species of sea turtles and shrimp fishing and other activities in the waters of the United States, Mexico and other nations of the world; and

(viii) the need for improved cooperation among departments, agencies and entities of Federal and State government, the need for improved cooperation with other nations and the need for treaties or international agreements on a bilateral or multilateral basis.

(7) MODIFICATION OF REGULATIONS.—For good cause, the Secretary may modify the regulations promulgated on June 29, 1987, relating to sea turtle conservation, in whole or part, as the Secretary deems advisable.

(8) SECRETARY AND EDUCATIONAL EFFORTS.—The Secretary shall undertake an educational effort among shrimp fishermen, either directly or by contract with competent persons or entities, to instruct fishermen in the usage of the turtle excluder device or any other device which might be imposed upon such fishermen;

(9) SEA TURTLE COORDINATION.—In order to coordinate the protection, conservation, reproductive, educational and recovery efforts with respect to each of the relevant species of sea turtles in accordance with existing law, the National Marine Fisheries Service shall designate an individual as Sea Turtle Coordinator to establish and carry out an effective, long-term sea turtle recovery program.

(10) PURPOSE OF THIS SECTION.—Section 8 is intended to assist the Secretary in making recommendations and in carrying out his duties under law, including the Endangered Species Act (16 U.S.C. 1531 et seq.), and nothing herein affects, modifies or alters the Secretary's powers or responsibilities to review, determine or redetermine, at any time, his obligations under law.

(11) DEFINITIONS.—For the purposes of this section, the terms:

(i) "relevant species of sea turtles" means the Kemp's ridley sea turtle, United States breeding populations of the loggerhead, the leatherback, and the green sea turtle, and other significant breeding populations of the loggerhead, the leatherback and the green sea turtle;

(ii) "status" means whether a given species of turtle is endangered, threatened or recovered;

(iii) "size" means the size of a given species of sea turtle; and

(iv) "age and sex structure" shall be considered to mean the distribution of juveniles, subadults and adults within a given species or population of sea turtles, and males and females within a given species or population of sea turtles.

(c) AUTHORIZATION OF APPROPRIATIONS.—There are authorized to be appropriated to the Department of Commerce $1,500,000 through fiscal year 1989 to carry out this section.

Appendix B

Interim Report

October 3, 1989

The Honorable Robert A. Mosbacher
Secretary of Commerce
Department of Commerce
14th Street and Constitution Avenue N.W.
Washington, D.C. 20230

Dear Mr. Secretary:

The National Research Council's Committee on Sea Turtle Conservation (see attached roster), convened at the request of the National Oceanic and Atmospheric Administration (NOAA) pursuant to Section 1008(b) of the Endangered Species Act Amendments of 1988 (ESAA), is pleased to provide this interim report on the Kemp's ridley sea turtle. As you know, Section 1008(b) of the ESAA specified an independent review by the National Academy of Sciences regarding scientific information on the biology and conservation of five species of sea turtles, all of which are classified as threatened or endangered. The statute stipulated that "in the event that the independent review cannot be completed by April 1, 1989, then the panel shall give priority to completing the independent review as it applies to the Kemp's ridley sea turtle and submitting same to the Secretary by that date, or as expeditiously as possible, and thereafter shall complete as expeditiously as possible the remaining work of the independent review."

Inasmuch as the Committee on Sea Turtle Conservation began its deliberations at its first meeting on May 4-5, 1989, pursuant to contract No. 50DGNC 9 00080 with NOAA, the review of all five sea turtle species has just begun. This interim report on the Kemp's ridley sea turtle summarizes the current knowledge on geographic distribution, population trends, causes of mortality, and protection measures. Although the committee's evaluation of the causes of mortality is incomplete at this time, the committee considers it important to issue this interim report now to communicate the seriousness of the Kemp's ridley's status. Important new literature on the biology and conservation of the Kemp's ridley, such as the biological synopsis by Márquez M. (1989, in review), a symposium proceedings edited by Caillouet and Landry (1989, in press), and a report on the status of the species by Ross et al. (Sept. 1989) will be available for interpretation and analysis in time for the committee's final report, which is scheduled for completion in February 1990. The final report will evaluate the biology, causes of mortality, and conservation of all five species of sea turtles.

BACKGROUND

The Kemp's ridley sea turtle (*Lepidochelys kempi* (Garman)) was first listed in the *Federal Register* as an endangered species on December 2, 1970, and its endangered status was reaffirmed in 1985 by the National Marine Fisheries Service (NMFS) on the basis of population figures appearing in many scientific publications (NOAA, 1985). Since 1947, when the first nesting Kemp's ridley turtles were discovered on a remote beach near Rancho Nuevo, Tamaulipas, Mexico, many data on the life history and population status of the species have been collected. So compelling are the data that the status of the species has never been seriously questioned since it was listed as endangered in 1970. Almost all the world's adult females of this species nest near Rancho Nuevo, and known habitats for the developing young Kemp's ridleys include both inshore and offshore waters of the Gulf of Mexico and coastal zones of the southeastern United States (especially Florida and Georgia).

GEOGRAPHIC DISTRIBUTION

The distribution (i.e., feeding habitats for juvenile stages and adults) of the Kemp's ridley is more restricted than that of any other sea turtle species. Preferred habitats by life stage have been identified by Pritchard (1969), Brongersma (1972), Pritchard and Márquez M. (1973), and others. Adult Kemp's ridleys are almost completely confined to the western Gulf of Mexico and very rarely appear east of Alabama in the northern Gulf. Adults concentrate to feed near the Louisiana coast in the North and southeastward off the coast of Campeche (Mexico). Juveniles appear almost entirely in U.S. waters of the Gulf of Mexico and along the southeastern Atlantic coast. However, lesser concentrations appear in various protected waters, including Chesapeake Bay, as far north as Long Island Sound (New York and Connecticut) and Vineyard Sound (Massachusetts), and rare strandings have been reported in western Europe, Malta, and Morocco (Brongersma, 1972; Brongersma and Carr, 1983; Manzella et al., 1988).

INDEX OF POPULATION SIZE

As in other sea turtle species, each female Kemp's ridley lays many eggs, about 105 eggs per nest (R. Márquez M., Instituto Nacional de Pesca, personal communication, 1989); almost all nesting activity occurs in April, May, and June. After an incubation period of 46-54 days, the overall population receives a short-lived "pulse" of many thousands of hatchlings from June through August. Once at sea, the young turtles are uncountable; in fact, available methods are inadequate for counting individuals of immature stages, adult males, or nonbreeding adult females. Sex ratios of adults in the wild are unknown. However, because most adult females typically nest annually, and since the species concentrates almost all its reproductive effort on the single beach at Rancho Nuevo, nesting females are easier to count than those of other sea turtle species. Because tagging studies have shown that some females might nest as many as three times each season (Márquez M. et al., 1981), the number of nests can exceed the number of nesting females in a given season. Thus, a reasonable estimate of the number of nesting females is obtained by dividing the number of nests by a factor that is somewhere between 1 and 3.

The number of nesting females each year is currently the best available index of population size. One disadvantage of this index is that the effectiveness of any conservation program designed to increase the survival of eggs and hatchlings might not be measurable for many years, because it is not known how long wild Kemp's ridleys require to attain sexual maturity. In captivity, maturity has been attained in as little as 4 years (Wood and Wood, 1984); in the wild, it might take as long as 10 to 15 years, based on results of studies done on other species. Studies of other sea turtle species in the wild have indicated not only slow growth to maturity, but also extreme variation in growth rate between individuals (Limpus, 1985; Bjorndal and Bolten, 1988). Thus, in the absence of an effective tagging program, we do not know how to relate changes in the numbers of nesting females to the number of hatchlings that might ultimately be recruited to the breeding population. Also, we do not know when such recruitment will occur.

LONG-TERM POPULATION TRENDS

In 1947, an estimated 40,000 female Kemp's ridleys were observed nesting during a single day at Rancho Nuevo (Carr, 1963; Hildebrand, 1963), as judged from a motion picture taken by an amateur photographer. For the 18 years following 1947, data are lacking on the status of the nesting colony near Rancho Nuevo. By 1966, the nesting assemblages, or "arribadas" (aggregations of nesting females at a given place at a given time), were much smaller; about 1,300 females nested on May 31, 1966 (Chávez et al., 1967). Since 1966, the Mexican government, working with the Estacion de Biologia Pesquera in Tampico and several other agencies, has maintained a presence on the beach at Rancho Nuevo throughout each nesting season. Personnel have included government turtle biologists, fisheries inspectors, and armed, uniformed Mexican marines. From 1967 to 1972, a few arribadas as large as 2,000-2,500 turtles were seen (Pritchard and Márquez M., 1973). Archival photographs, probably from 1968, show many hundreds of nesting females on the beach. Overall, from 1947 to 1970, sizes of the largest arribadas on the Rancho Nuevo beach have dramatically declined (Figure 1).

Since 1978, nests have been counted on the beach by a binational team of Mexican and U.S. scientists working with the U.S. Fish and Wildlife Service (USFWS). The number of female turtles observed on that beach has declined through 1988 to 500-700 for the entire season (USFWS Annual Reports, Albuquerque office, 1978-1988). From intensive observations and captures of nesting females, Márquez M. et al. (1981) estimated that each female nests 1.3 times each year. Dividing by this factor to convert nest counts to total nesting females, we estimate that 711 nested in 1978, 655 in 1988, and, for the period 1978-1988, an average of 626 female ridleys nested each year at Rancho Nuevo. Improved coverage of the nesting beach by biologists during the 1980s indicates that the 1.3 value is a low estimate and will need to be revised upward to 1.4 or 1.55

(Márquez M., 1989; NMFS, 1989). (Regardless of the factor used, the form of the trend in Figure 1 remains the same.) Thus, the 626 estimate is probably an overestimate, and the actual number of nesting turtles is less. Nevertheless, even the 626 annual average during the decade represents only 1.5% of the minimal estimate for the 1947 population (see also Frazer, 1986; NMFS, 1989). For the 1978-1988 period, which included some annual fluctuations to be expected in sea turtle populations (Richardson, 1982), the estimated number of nesting females declined significantly (linear regression, $P < .05$), at a rate of approximately 14 females per year, or 140 females over a decade (Figure 2).

Efforts have been made by aerial surveys, foot patrols, and interviews with local residents to determine whether major nesting occurs away from Rancho Nuevo, but no additional major nesting site has been found.[1] In most years, one or two individuals have nested on Padre Island, Texas, and a few dozen near Tecolutla, Veracruz (Márquez M., 1989). The nesting record farthest from Rancho Nuevo is from the vicinity of Isla Aguada, Campeche (NMFS, 1989), apart from a recent, single nesting near St. Petersburg, Florida, in 1989 and an attempted nesting in Broward County, Florida, in 1989.

CAUSES OF MORTALITY

At different stages of their life cycle, Kemp's ridleys can be adversely affected by a number of activities and substances. These potentially include severe changes in weather and associated conditions (including high tides and waves) at nesting beaches; cold-stunning; human and non-human depredation of eggs in nests; predation of hatchlings and/or older turtles by crabs, birds, fish, and mammals; industrial pollutants; diseases; exploratory oil and gas drilling; dredging; and incidental capture in shrimp nets and other fishing gear (Coston-Clements and Hoss, 1983). Several of these factors, including severe weather and industrial pollution, are unverified causes of mortality, and natural predation by birds, fish, or mammals is a factor with which the species has coexisted throughout its evolutionary history. The effects of natural predation (e.g., increased

[1] In 1988, Hurricane Gilbert reportedly rearranged the historical nesting beach at Rancho Nuevo. Some nesting turtles were located in 1989 approximately 25 km north of the Rancho Nuevo beach survey camp, an area not patrolled from the ground in previous years. Full details on this observation, as well as complete numbers of nests for 1989, have not yet been received from the collaborative team of Mexican Fisheries Department and the U.S. Fish and Wildlife Service (Jack Woody, USFWS, personal communication, 5 September 1989).

Figure 1. Number of nesting females at Rancho Nuevo in largest arribada.

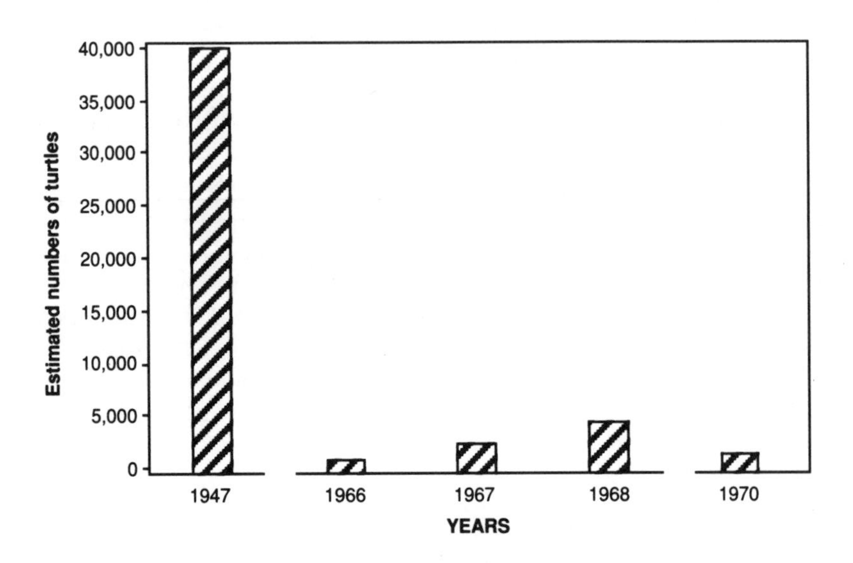

Figure 2. Number of nesting females at Rancho Nuevo, estimated from numbers of nests found. The linear regression line has a slope of -14.25 and is significant ($p < .05$).

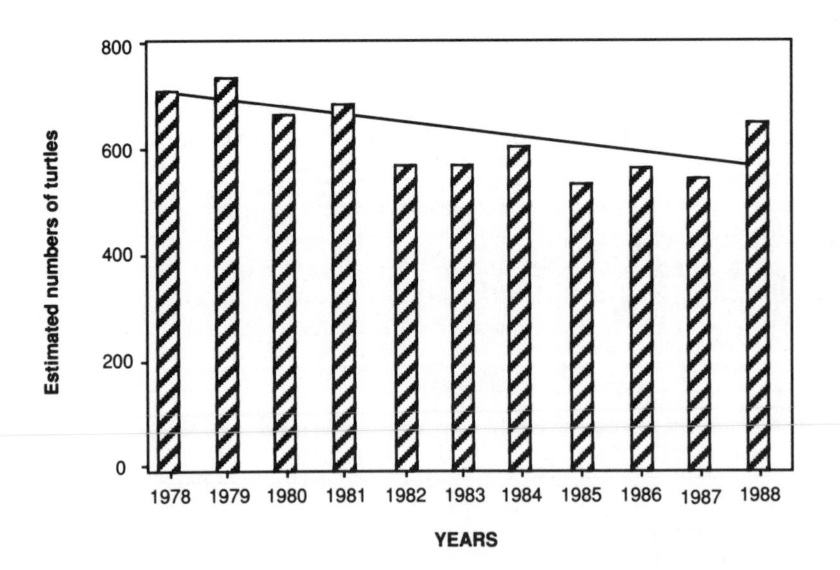

mortality rate), however, might become more severe at low population densities.

One particular source of mortality, the incidental capture of sea turtles in fishing gear, has been well documented. For example, recently Murphy and Hopkins-Murphy (1989) reviewed 78 papers on the incidental capture of all Atlantic sea turtle species in which various types of fishing gear were used. Shrimp-trawling was documented or implicated as a major source of mortality in 83% of these papers. Specifically regarding Kemp's ridleys, for many years the primary source of tag returns from females nesting at Rancho Nuevo (84% of 129 returns) was the accidental capture of turtles and subsequent reporting of tag numbers by helpful shrimpers (Pritchard and Márquez M., 1973; Márquez M., 1989). From January 1980 to March 31, 1989, the Sea Turtle Stranding and Salvage Network documented 976 stranded dead Kemp's ridleys on the beaches of prime shrimp-trawling areas between North Carolina and Texas (Anonymous, 1987; Schroeder and Warner, 1988; Warner, 1988; Teas, 1989). This indication of continuing high mortality is considered by turtle population biologists as a distinct threat to the survival of the species. Although the committee has not yet evaluated the relative impacts of all potential mortality factors affecting the Kemp's ridley, incidental entrapment in fishing gear is clearly a major cause of mortality. An analysis of each mortality factor will be provided in the committee's final report.

PROTECTIVE MEASURES

Legal Protection

The Kemp's ridley has been legally protected in Mexico since 1966 and in the United States since 1973. In no other country does the species occur, except as an occasional waif (straggler), highly unlikely ever to breed. The species is listed as an Appendix I species by the Convention on International Trade in Endangered Species (CITES) and is thus prohibited in international commerce between, from, or to signatory countries. The International Union for the Conservation of Nature and Natural Resources also considers the species endangered. Potential benefits of protective measures will be reviewed in the committee's final report.

Protection at the Nesting Colony

Since 1966, the Mexican government (initially the Subsecretaria de Industria y Comercio, now the Instituto Nacional de Pesca) has maintained a seasonal camp at Rancho Nuevo to protect nesting turtles and their eggs. Eggs are quickly moved to a guarded hatchery to prevent natural and human predation. From 1966 to 1977, an average of 23,000 hatchlings was released each year at Rancho Nuevo (Márquez M., 1989). Although many nests were raided during these years, very few adults were killed. Since 1978, the beach effort has been binational (USFWS plus the Instituto Nacional de Pesca), and only about 10% of the nests have been lost each year. The number of eggs moved to a hatchery has ranged from 98,211 (1979) to 65,357 (1986). Hatching percentage has averaged 61%, ranging from 53% in 1983 to 75% in 1985 and 1986. An average of 48,633 hatchlings has been released each year since 1978 (annual project reports from USFWS Albuquerque office). Based on a comparison with 1966-67, the results of these beach efforts clearly show that protective measures at the nesting beach have improved recruitment at the hatchling level in the population.

Turtle Excluder Devices (TEDs)

Over the last 10 years, NMFS has developed and tested a device that fits into the throat of a bottom (shrimp) trawl to exclude sea turtles. Several other devices that work on similar principles have also been developed by industry and tested by NMFS for turtle exclusion. After extensive debate, public hearings, and legal challenges, NMFS promulgated a regulation on June 29, 1987, requiring TED use in U.S. Atlantic and Gulf Coast waters (offshore areas by May 1, 1989; inshore areas by May 1, 1990; Conner, 1987). In 1988, South Carolina established a regulation requiring the use of TEDs by shrimp trawlers in state waters at all times. In early 1989, Florida passed a similar regulation (Rule No. 46-31.002, governor and cabinet) requiring TED use in state waters along the Atlantic Coast north of the Brevard-Volusia County line. Regulations now include provisions for certification of additional TED designs originating in the shrimping industry. The committee notes that TED regulations have been changing rapidly.

HEADSTARTING

Headstarting is an experimental program, the actual benefits of which have yet to be evaluated. In headstarting, hatchling turtles are raised in captivity for several months before release as a supplement to the beach protective efforts. Personnel of the NMFS and other agencies have raised hatchling ridleys from about 2,000 eggs donated by Mexico each year. At about 10 months of age, the turtles have been released, at first into Florida waters where juvenile Kemp's ridleys are known to appear, but in recent years mostly off Padre Island, Texas. A total of 12,422 turtles was released from 1978 to 1986 (Manzella et al., 1988).

Conclusions

The Kemp's ridley sea turtle, nesting almost exclusively on a single Mexican beach near Rancho Nuevo, is restricted largely to the northern and southern Gulf of Mexico and the Atlantic seaboard of the United States. The most reliable index for estimating population size in this species is the annual number of nesting adult females at Rancho Nuevo. At that location, the number of nesting females has decreased from an estimated 40,000 (in a single day) in 1947 to an estimated 655 in the 1988 season, a decline that clearly signals a serious threat to the existence of the species. Protection of the nesting turtles, nests, and eggs on the Rancho Nuevo beach has resulted in increased numbers of hatchlings in recent years. Causes of mortality have been identified, but an analysis of their relative impacts must be deferred to the committee's final report.

We hope that this interim report is useful to you.

Sincerely,

John J. Magnuson
Chairman
Committee on Sea Turtle Conservation

cc: Dr. John A. Knauss, NOAA Administrator
David Cottingham, NOAA, Contract Officers Technical Representative

REFERENCES

Anonymous. 1987. Final Supplement to the Final Environmental Impact Statement on Listing and Protecting the Green Sea Turtle, Loggerhead Sea Turtle, and Pacific Ridley Sea Turtle Under the Endangered Species Act of 1973. St. Petersburg, Fla.: National Oceanic and Atmospheric Administration/National Marine Fisheries Services. 59 pp.

Bjorndal, K. A., and A. B. Bolten. 1988. Growth rates of immature green turtles, *Chelonia mydas,* on feeding grounds in the southern Bahamas. Copeia 1988:555-564.

Brongersma, L. D. 1972. European Atlantic Turtles. Uitgegeven door het Rijksmuseum van Natuurlicke Historie te Leiden 121:1-318.

Brongersma, L. D., and A. F. Carr. 1983. *Lepidochelys kempi* (Garman) from Malta. Proceedings Koninklicke Nederlandse Acadamie van Wetenschappen Series C 86(4):445-454.

Caillouet, C. W., Jr., and A. M. Landry, Jr. 1989. Proceedings of the First International Symposium on Kemp's Ridley Sea Turtle Biology, Conservation and Management. Texas A & M Sea Grant Program No. 89-105, Galveston, Tex. (In press).

Carr, A. F. 1963. Panspecific reproductive convergence in *Lepidochelys kempi*. Ergebnisse der Biologie 26:298-303.

Cáhvez, H., M. Contreras, and T. P. E. Hernandez. 1967. Apectos biologicos y proteccion de la Tortuga Lora, *Lepidochelys kempi* (Garman), en la Costa de Tamaulipas, Mexico. Instituto Nacional Investigaciones Biologico-Pesqueras, Mexico 17:1-39.

Conner, D. K. 1987. Turtles, trawlers, and TEDS: what happens when the Endangered Species Act conflicts with fishermen's interests. Water Log (Coastal and Marine Law Research Program, University of Mississippi) 7(4): 3-27.

Coston-Clements, L., and D. E. Hoss. 1983. Synopsis of Data on the Impact of Habitat Alteration on Sea Turtles around the Southeastern United States. NOAA Technical Memorandum National Oceanic and Atmospheric Administraton/National Marine Fisheries Service, Southeast Fisheries Center. 117pp. National Oceanic and Atmospheric Administration. 57 pp.

Frazer, N. B. 1986. Kemp's decline: Special alarm or general concern? Marine Turtle Newsletter 37:5-7.

Hildebrand, H. H. 1963. Hallazgo del area de anidacion de la Tortuga Marina "Lora" *Lepidochelys kempi* (Garman), en la Costa Occidental del Golfo de Mexico. (Rept., Chel.). Ciencia XXII (4):105-112.

Limpus, C. J. 1985. A Study of the Loggerhead Sea Turtle, *Caretta caretta,* in Eastern Australia. Ph.D. dissertation, University of Queensland, St. Lucia, Australia.

Manzella, S. A., C. W. Caillouet, Jr., and C. T. Fontaine. 1988. Kemp's ridley, *Lepidochelys kempi,* sea turtle head start tag recoveries: Distribution, habitat and method of recovery. Mar. Fish. Rev. 50:33-42.

Márquez M., R. 1989. Synopsis of the Biological Data on the Kemp's Ridley Sea Turtle *Lepidochelys kempi* (Garman, 1880). Unpublished manuscript, in review.

Márquez M., R., A. Villaneuva O., and M. Sanchez P. 1981. The population of Kemp's ridley sea turtle in the Gulf of Mexico, *Lepidochelys kempi*. Pp. 159-164 in Biology and Conservation of Sea Turtles, K. Bjorndal, ed. Washington, D.C.: Smithsonian Institution Press.

Murphy, T. M., and S. R. Hopkins-Murphy. 1989. Sea turtle & Shrimp Fishing Interactions: A Summary and Critique of Relevant Information. Washington, D.C.: Center for Marine Conservation. 52 pp.

NMFS (National Marine Fisheries Service). 1989. Draft Report of the International Committee for the Recovery Plan for Kemp's Ridley Sea Turtle. David Owens (Texas A & M University), Chairman. St. Petersburg, Fla.

NOAA (National Oceanic and Atmospheric Administration). 1985. Five-Year Status Reviews of Sea Turtles Listed Under the Endangered Species Act of 1973. Prepared by Andreas Mager, Jr., National Marine Fisheries Service, St. Petersburg, Fla.

Pritchard, P. C. H. 1969. Studies of the Systematics and Reproductive Cycles of the Genus Lepidochelys. Ph.D. Dissertation, University of Florida, Gainesville, Fla. 197 pp.

Pritchard, P. C. H., and R. Márquez M. 1973. Kemp's Ridley Turtle or Atlantic Ridley, *Lepidochelys kempi*. IUCN Monograph No. 2: Marine Turtle Series. Morges, Switzerland: International Union for the Conservation of Nature and Natural Resources. 30 pp.

Richardson, J. I. 1982. A population model for adult female loggerhead sea turtles (*Caretta caretta*) nesting in Georgia. Ph. D. Dissertation, University of Georgia, Athens, Ga.

Ross, J. P., S. Beavers, D. Mundell, and M. Airth-Kindree. 1989. The Status of Kemp's Ridley. September 1989. Washington, D. C.: Center for Marine Conservation. 51 pp.

Schroeder, B. A., and A. A. Warner. 1988. 1987 Annual Report of the Sea Turtle Stranding and Salvage Network: Atlantic and Gulf Coast States of the United States. January-December 1987. Coastal Resources Division Contract No. CRD-87/88-28. Miami, Fla.: National Oceanic and Atmospheric Administration/National Marine Fisheries Service, Southeast Fisheries Center. 45 pp.

Teas, W. 1989. 1989 First Quarter Report of the Sea Turtle Stranding and Salvage Network. Atlantic and Gulf Coasts of the United States. January-March 1989. National Oceanic and Atmospheric Administration/National Marine Fisheries Service, Southeast Fisheries Center. Coastal Resources Division (in press).

Warner, A. A. 1988. 1988 Third Quarter Report of the Sea Turtle Stranding and Salvage Network Atlantic and Gulf Coasts of the United States. January- September 1988. Coastal Resources Division Contract No. CRD-88/89-01. National Oceanic and Atmospheric Administration/National Marine Fisheries Service, Southeast Fisheries Center. 22 pp.

Wood, J. R., and F. E. Wood. 1984. Captive breeding of the Kemp's ridley. Marine Turtle Newsletter 29:12.

COMMITTEE ON SEA TURTLE CONSERVATION

John J. Magnuson, University of Wisconsin (Chairman), Madison, Wisconsin
Karen Bjorndal, University of Florida, Gainesville, Florida
William D. DuPaul, Virginia Institute of Marine Science, Gloucester Pt., Virginia
Gary L. Graham, Texas Agricultural Extension Service, Freeport, Texas
David W. Owens, Texas A & M University, College Station, Texas
Charles H. Peterson, University of North Carolina, Morehead City, North Carolina
Peter C. H. Pritchard, Florida Audubon Society, Maitland, Florida
James I. Richardson, University of Georgia, Athens, Georgia
Gary E. Saul, Southwest Texas State University, Austin, Texas
Charles W. West, Nor'Eastern Trawl Systems Inc., Bainbridge Is., Washington

Staff

David Johnston, Project Director
Linda B. Kegley, Project Assistant

BOARD ON ENVIRONMENTAL STUDIES AND TOXICOLOGY

Gilbert S. Omenn, University of Washington (Chairman), Seattle, Washington
Frederick R. Anderson, American University, Washington, D.C.
John Bailar, McGill University School of Medicine, Montreal, Quebec
Lawrence W. Barnthouse, Oak Ridge National Laboratory, Oak Ridge, Tennessee
David Bates, University of British Columbia, Vancouver, British Columbia
Joanna Burger, Rutgers University, Piscataway, New Jersey
Yorman Cohen, University of California, Los Angeles, California
John L. Emmerson, Eli Lilly & Company, Greenfield, Indiana
Robert L. Harness, Monsanto Agricultural Company, St. Louis, Missouri
Paul J. Lioy, UMDNJ-Robert Wood Johnson Medical School, Piscataway, New Jersey
Jane Lubchenco, Oregon State University, Corvallis, Oregon
Donald Mattison, University of Arkansas, Little Rock, Arkansas

Duncan T. Patten, Arizona State University, Tempe, Arizona
Nathaniel Reed, Hobe Sound, Florida
William H. Rodgers, University of Washington, Seattle, Washington
F. Sherwood Rowland, University of California, Irvine, California
Liane B. Russell, Oak Ridge National Laboratory, Oak Ridge, Tennessee
Milton Russell, University of Tennessee and Oak Ridge National Laboratory, Knoxville, Tennessee
John H. Seinfeld, California Institute of Technology, Pasadena, California
I. Glenn Sipes, University of Arizona, Tucson, Arizona

Staff

James J. Reisa, Acting Director
Karen Hulebak, Program Director for Exposure Assessment and Risk Reduction
David Policansky, Program Director for Applied Ecology and Natural Resources
Richard Thomas, Program Director for Human Toxicology and Risk Assessment

BOARD ON BIOLOGY

Francisco J. Ayala (Chairman), University of California, Irvine
Nina V. Fedroff, Carnegie Institution of Washington, Baltimore, Maryland
Timothy H. Goldsmith, Yale University, New Haven, Connecticut
Ralph W. F. Hardy, BioTechnica/Boyce Thompson Institute, Ithaca, New York
Ernest G. Jaworski, Monsanto Company, St. Louis, Missouri
Simon A. Levin, Cornell University, Ithaca, New York
Harold A. Mooney, Stanford University, Stanford, California
Harold J. Morowitz, George Mason University, Fairfax, Virginia
William E. Paul, National Institutes of Health, Bethesda, Maryland
David D. Sabatini, New York University, New York
Malcolm S. Steinberg, Princeton University, Princton, New Jersey
David B. Wake, University of California, Berkeley
Bruce M. Alberts (ex-officio), University of California, San Francisco

Oskar R. Zaborsky, Director

COMMISSION ON LIFE SCIENCES

Appendix C
Illustrations of Turtle Excluder Devices

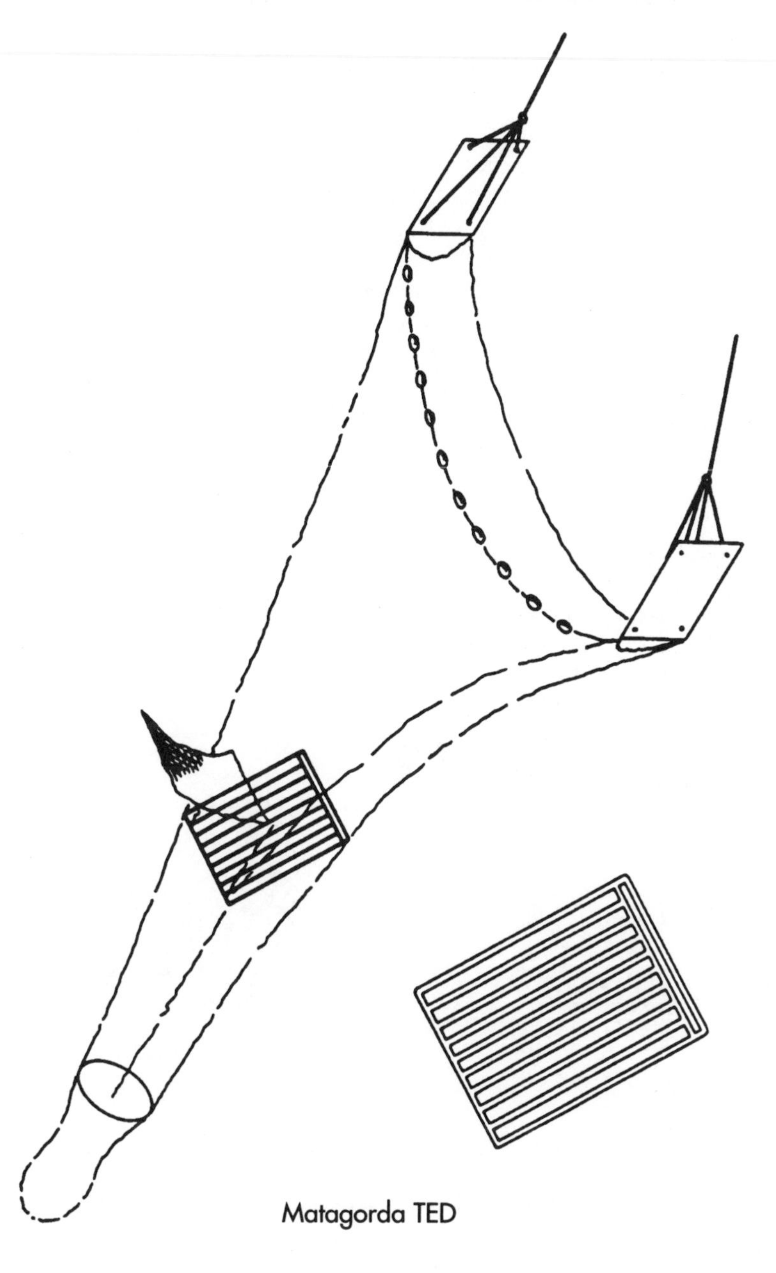

Matagorda TED

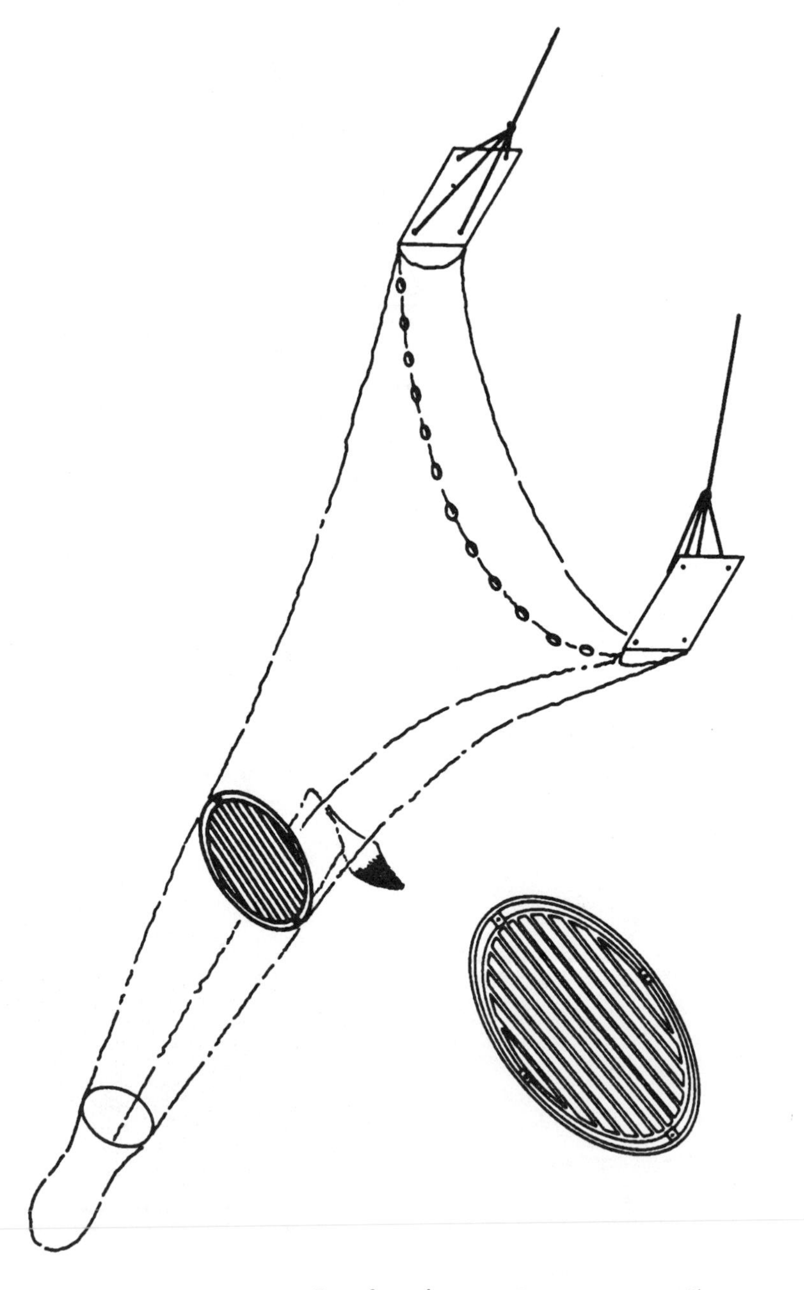

Georgia TED (typically referred to as "Georgia jumper")

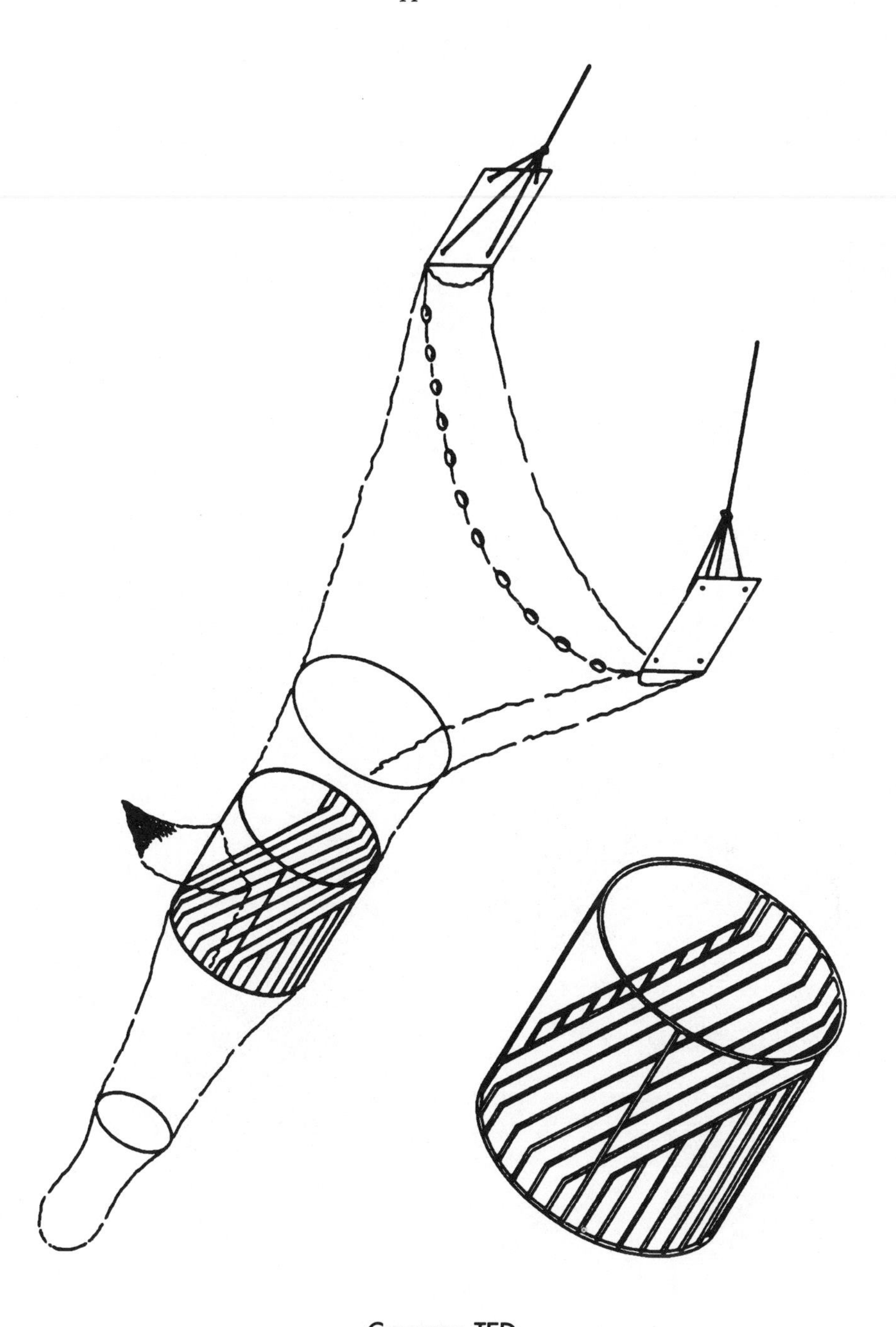

Cameron TED

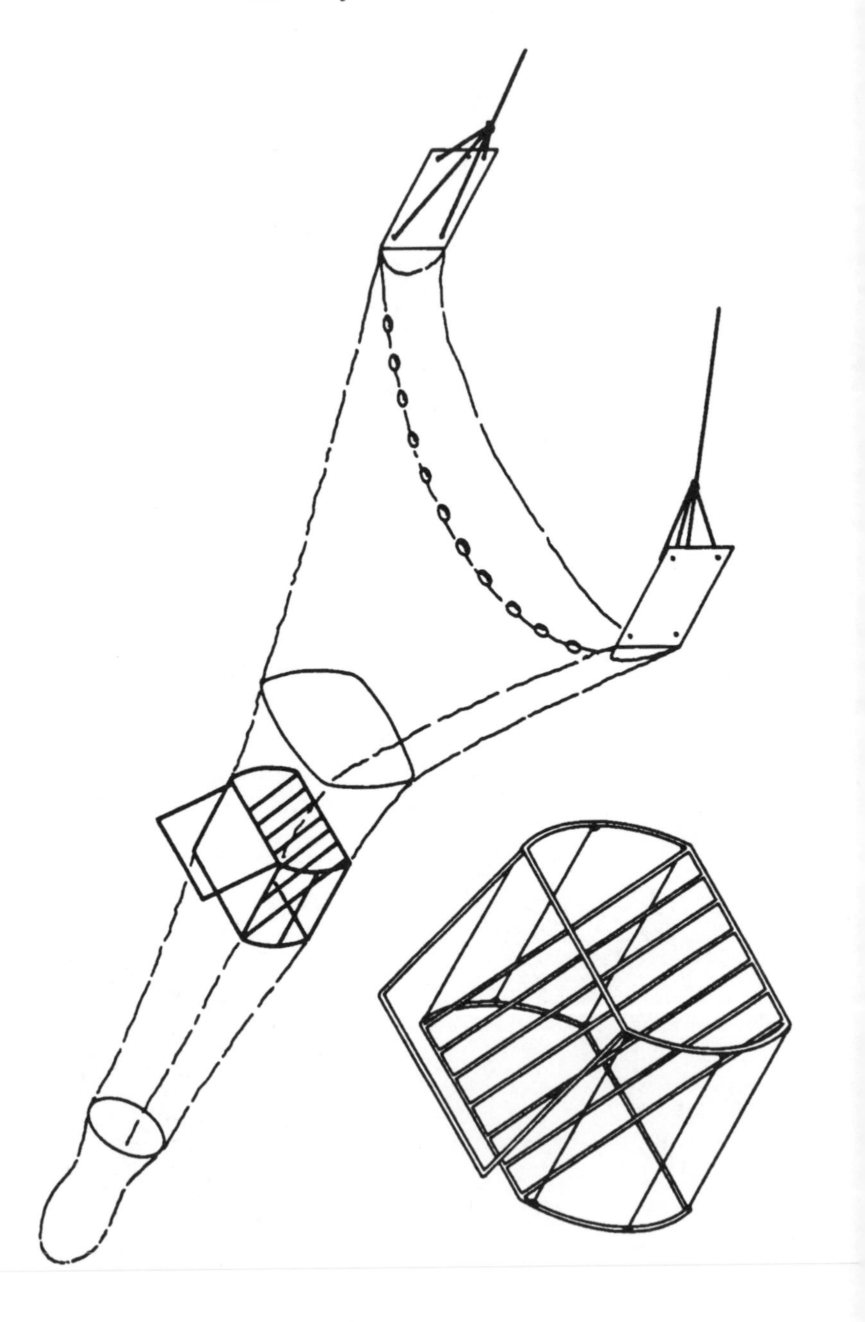

NMFS TED

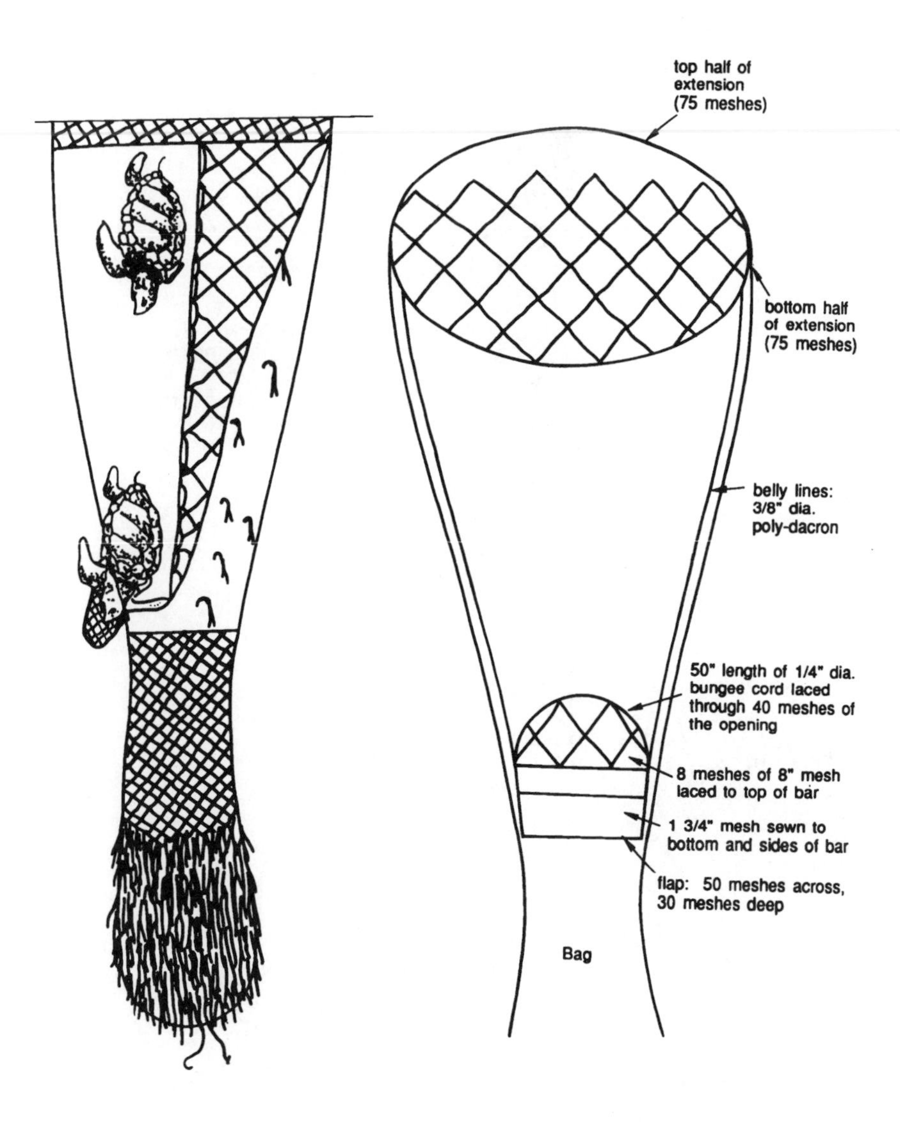

Parrish Soft TED

(Reprinted with permission of N.C. Sea Grant College Program and the Georgia Sea Grant College Program.)

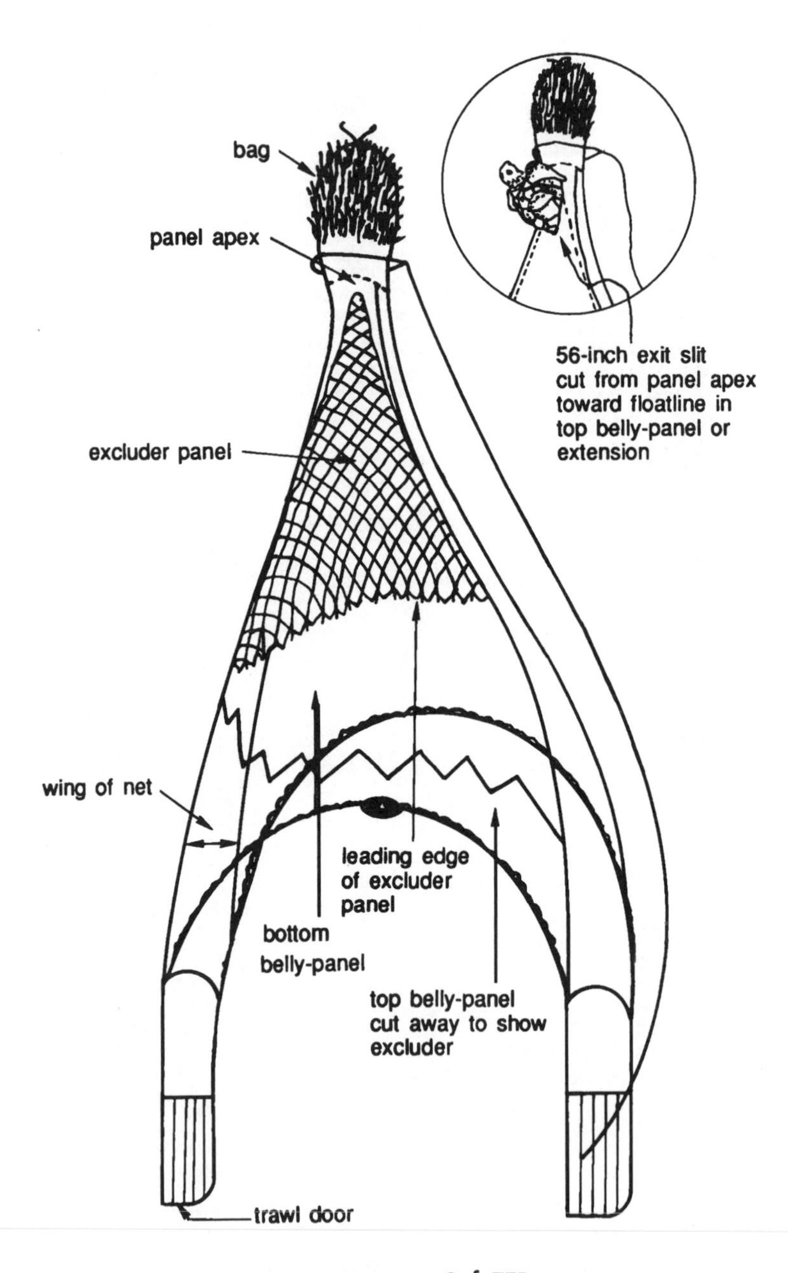

Morrison Soft TED
(Reprinted with permission of the Georgia Sea Grant College Program.)

Appendix D

Aerial Survey Data of Sea Turtles in Fishing Zones

Compiled by the committee from Winn (1982), Thompson (1984), Loehoefener (personal communication, NMFS, 1989), and Thompson (personal communication, NMFS, 1989).

AERIAL SURVEY DATA FROM THOMPSON DENSITIES SIGHTED PER 10,000KM2. FROM FLORIDA TO THE MISSISSIPPI RIVER I WEIGHTED OFFSHORE AREA BY 3 AND INSHORE AREAS BY 1 TC

MOSTLY LOGGERHEAD | LOGGERHEAD

STATE	TX	TX	TX	TX	LA	LA	LA	LA	LA	LA/MS	MS/AL	AL/GFL	GFL	GFL	GFL	GFL	GFL	GFL	GFL	KEYS	KEYS	AFL	AFL
ZONE	21	20	19	18	17	16	15	14	13	12	11	10	9	8	7	6	5	4	3	1	24	25	26
SEASON																							
WIN	1	1	1	1	10	10	10	10	10									373	373	373	58	58	157
SPR	7	7	7	7	0	0	0	0	0	135	135	135	135	135							981	981	1975
SUM	2	2	2	2	5	5	5	5	5	52	52	52	52	52	92	92	92	97	97	97	463	463	5242
FALL	12	12	12	12	2	2	2	2	2	246	246	246	246	246	147	147	147	152	152	152	416	416	588

CALCULATE A MEAN DENSITY. GULF OF MAINE = MEAN OF GULF AND GEORGE'S BANK.

STATE	AFL	AFL	AFL	AFL/GA	GA	SC	SC/NC	NC	NC	NC/VA	VA	MD/DE	NJ	NJ/NY	T/RI/MA	MA/NH	ME
ZONE	27	28	29	30	31	32	33	34	35	36	37	38	39	40	41	42	43
SEASON																	
WIN	157	157	163	976	917	1244	390		0	0	0	0	0	0	0	0	0
SPR	1975	1975	7863	1954	524	891	2502	1398	98	98	98	98	98	98	98	1	1
SUM	5242	5242	7863	658	530	475	1491	379	525	525	525	525	525	525	525	4	4
FALL	588	588	510	349	213	545	2426	1820	161	161	161	161	161	161	161	0	0

MOSTLY LOGGERHEADS, AERIAL SURVEY DATA MODIFIED FROM THOMPSON FOR THE ATLANTIC AND BY LOHOEFENER FOR THE GULF, MOSTLY LOGGERHEADS

STATE	TX	TX	TX	TX	LA	LA	LA	LA	LA	LA/MS	MS/AL	AL/GFL	GFL	GFL	GFL	GFL	GFL	GFL	GFL	KEYS	KEYS	AFL	AFL
ZONE	21	20	19	18	17	16	15	14	13	12	11	10	9	8	7	6	5	4	3	1	24	25	26
SEASON																							
WIN																		64	64	64	58	58	157
SPR	34	34	23	23	4	4	4	4	4	4	30										981	981	1975
SUM															33						463	463	5242
FALL	16	16	14	14	5	5	5	5	5	5	19				35	32	32	113	113	113	416	416	588

									XXX	LEATHERBACKS from Thompson thesis; no. sighted per 10,000km2									
STATE	AFL	AFL	AFL	AFL/GA	GA	SC	SC/NC	NC	XXX		NC	NC/VA	VA	MD/DE	NJ	NJ/NY	T/RI/MA	MA/NH	ME
ZONE	27	28	29	30	31	32	33	34	XXX		35	36	37	38	39	40	41	42	43
SEASON									XXX										
WIN	157	157	163	976	917	1244	390		XXX	WIN	0	0	0	0	0	0	0	0	0
SPR	1975	1975	7863	1954	524	891	2502	1398	XXX	SPR	1	1	1	1	1	1	1	1	1
SUM	5242	5242	7863	658	530	475	1491	379	XXX	SUM	19	19	19	19	19	19	19	7	7
FALL	588	588	510	349	213	545	2426	1820	XXX	FALL	3	3	3	3	3	3	3	8	8

Appendix E

Sea Turtle Stranding Data

Compiled by the committee from STSSN data and unpublished raw NMFS data.

LOGGERHEAD

NOTES: 1987 STRANDINGS OFFSHORE WITHOUT HEADSTART TURTLES, unidentified included in total sea turtles

1987	TX	TX	TX	TX	LA	LA	LA	LA	LA	LA/MS	MS/AL	AL/GFL	GFL	GFL	GFL	GFL	GFL	GFL	GFL	KEYS	KEYS	AFL
ZONE	21	20	19	18	17	16	15	14	13	12	11	10	9	8	7	6	5	4	3	1	24	25
MONTH																						
JAN	0	0	0	0	1	0	0	0	0	0	0	0	0	0	0	0	0	2	0	0	0	0
FEB	3	2	1	0	0	0	0	0	0	0	0	0	0	0	0	0	0	3	0	0	0	0
MAR	4	1	1	0	1	0	0	0	0	0	0	0	0	0	0	0	10	18	0	0	0	0
APR	12	27	9	2	2	0	0	0	0	0	0	1	0	1	1	0	11	10	0	0	0	0
MAY	4	14	6	4	3	0	0	0	1	0	11	11	0	2	2	0	7	11	0	0	0	1
JUN	0	3	2	2	4	0	0	3	0	0	5	24	0	2	0	0	5	5	0	0	0	1
JUL	2	5	5	4	1	0	0	0	0	0	1	0	0	1	0	0	2	3	0	1	0	1
AUG	1	2	1	3	0	0	1	0	0	0	0	0	1	0	0	0	3	2	0	1	0	0
SEP	0	3	3	1	2	0	0	0	0	0	0	0	0	1	0	0	0	1	0	0	1	0
OCT	1	2	2	1	0	0	0	0	0	0	2	0	0	0	0	0	0	2	0	0	0	0
NOV	0	1	1	0	0	0	0	0	0	0	0	0	0	0	0	0	1	0	0	0	0	0
DEC	3	2	0	2	0	0	0	0	0	0	0	0	0	0	0	0	0	3	0	0	0	2
TOTAL	30	62	31	19	14	0	1	3	1	0	19	36	1	7	3	0	39	60	0	2	1	5

1987	AFL	AFL	AFL	AFL	AFL/GA	GA	SC	SC/NC	NC	NC	NC/VA	VA	MD/DE	NJ	NJ/NY	T/RI/MA	MA/NH	ME		P.RICO
ZONE	26	27	28	29	30	31	32	33	34	35	36	37	38	39	40	41	42	43	TOTAL	99
MONTH																				
JAN	1	0	1	1	1	0	0	0	0	11	0	0	0	0	0	0	0	0	18	0
FEB	5	1	6	1	0	0	0	0	0	2	0	0	0	0	0	0	0	0	24	0
MAR	2	3	3	2	0	0	0	0	5	1	0	0	0	0	0	0	0	0	51	0
APR	2	5	2	5	7	1	0	0	0	0	0	0	0	0	0	0	0	0	98	0
MAY	3	13	7	6	56	74	20	23	12	1	2	0	0	0	0	0	0	0	294	0
JUN	0	6	10	40	50	84	125	12	22	2	14	3	0	0	2	0	0	0	426	9
JUL	1	4	9	32	36	24	44	22	12	0	3	0	1	3	3	0	0	0	220	0
AUG	1	6	6	8	39	10	10	11	12	3	3	5	1	6	5	0	0	0	141	0
SEP	0	2	4	11	29	6	13	13	2	0	4	0	4	4	5	0	0	0	109	0
OCT	1	1	0	3	6	8	3	0	0	3	6	0	3	3	1	0	0	0	48	0
NOV	1	2	2	11	20	4	0	1	1	1	1	0	0	0	1	0	0	0	48	0
DEC	1	4	0	8	10	1	0	0	1	8	0	0	0	0	0	0	0	0	45	0
TOTAL	18	47	50	128	254	212	215	82	67	32	33	8	9	16	17	0	0	0	1522	9

KEMP'S RIDLEY

1987	TX	TX	TX	TX	LA	LA	LA	LA	LA	LA/MS	MS/AL	AL/GFL	GFL	GFL	GFL	GFL	GFL	GFL	GFL	KEYS	KEYS	AFL
ZONE	21	20	19	18	17	16	15	14	13	12	11	10	9	8	7	6	5	4	3	1	24	25
MONTH																						
JAN	0	0	0	0	0	0	0	0	0	0	0	0	0	0	0	0	0	0	0	0	0	0
FEB	0	2	0	0	0	0	0	0	0	0	0	0	0	0	0	0	0	2	0	0	0	0
MAR	1	4	0	0	1	0	0	0	0	0	0	0	0	0	0	0	1	2	0	0	0	0
APR	0	6	3	1	0	0	0	0	0	0	0	0	0	0	0	0	0	1	0	0	0	0
MAY	0	0	0	1	7	0	0	2	1	0	0	0	0	1	0	0	1	0	0	0	0	0
JUN	0	0	1	0	0	0	0	1	1	0	0	0	0	0	0	0	0	0	0	0	0	0
JUL	0	3	0	0	3	0	0	0	0	0	0	0	0	0	0	0	0	0	0	0	0	0
AUG	0	1	2	4	2	0	0	0	0	0	0	0	0	0	1	0	0	0	0	0	0	0
SEP	0	0	1	4	0	0	0	0	2	0	0	0	0	0	0	0	0	0	0	0	0	0
OCT	0	0	4	2	0	0	0	0	0	0	4	0	0	0	0	0	0	0	0	0	0	0
NOV	0	1	0	2	0	0	0	0	0	0	0	0	0	0	0	0	1	0	0	0	0	0
DEC	0	3	4	0	0	0	0	1	0	0	0	0	0	0	0	0	0	0	0	0	0	0
TOTAL	1	20	15	14	13	0	0	4	4	0	4	0	0	1	1	0	3	5	0	0	0	0

1987	AFL	AFL	AFL	AFL	AFL/GA	GA	SC	SC/NC	NC	NC	NC/VA	VA	MD/DE	NJ	NJ/NY	T/RI/MA	MA/NH	ME		P.RICO
ZONE	26	27	28	29	30	31	32	33	34	35	36	37	38	39	40	41	42	43	TOTAL	99
MONTH																				
JAN	0	0	0	0	0	0	0	0	0	0	0	0	0	0	0	0	0	0	0	0
FEB	0	0	0	0	0	0	0	0	0	0	0	0	0	0	0	0	0	0	4	0
MAR	0	0	0	0	0	0	0	0	0	1	0	0	0	0	0	0	0	0	10	0
APR	0	0	0	0	0	0	0	0	0	0	0	0	0	0	0	0	0	0	11	0
MAY	0	0	0	0	0	1	0	1	0	0	0	0	0	0	0	0	0	0	15	0
JUN	0	0	1	0	2	3	4	1	0	0	1	0	0	0	0	0	0	0	15	0
JUL	0	0	0	0	1	1	0	1	1	0	0	0	0	0	0	0	0	0	10	0
AUG	0	0	0	0	0	2	1	0	0	0	0	0	0	0	0	0	0	0	13	0
SEP	0	0	0	0	0	1	1	1	0	0	0	0	0	0	0	1	0	0	11	0
OCT	0	0	0	0	1	0	0	1	0	0	2	0	0	0	1	0	0	0	15	0
NOV	0	0	0	4	7	2	0	0	0	0	1	0	0	0	0	0	0	0	18	0
DEC	0	0	0	4	0	0	0	0	1	6	0	0	0	0	0	0	0	0	19	0
TOTAL	0	0	1	8	11	10	6	5	2	7	4	0	0	0	1	1	0	0	141	0

GREEN 1987	TX	TX	TX	TX	LA	LA	LA	LA	LA	LA/MS	MS/AL	AL/GFL	GFL	GFL	GFL	GFL	GFL	GFL	GFL	KEYS	KEYS	AFL
ZONE	21	20	19	18	17	16	15	14	13	12	11	10	9	8	7	6	5	4	3	1	24	25
MONTH																						
JAN	0	0	0	0	0	0	0	0	0	0	0	0	0	0	0	1	3	0	0	0	1	0
FEB	0	0	0	0	0	0	0	0	0	0	0	0	0	0	0	0	1	0	1	1	1	0
MAR	0	0	0	0	0	0	0	0	0	0	0	0	0	0	0	0	1	0	0	1	2	1
APR	0	3	0	0	0	0	0	0	0	0	3	0	0	0	0	0	1	0	0	0	0	0
MAY	1	0	0	1	0	0	0	0	0	0	1	0	0	0	0	0	0	0	0	0	0	2
JUN	0	1	0	0	0	0	0	0	0	0	0	0	0	0	0	0	1	0	0	0	0	0
JUL	0	0	1	0	0	0	0	0	0	0	0	0	0	0	0	0	0	1	0	1	1	0
AUG	0	1	0	0	0	0	0	0	0	0	0	0	0	0	0	0	0	0	0	0	0	0
SEP	0	0	0	0	0	0	0	0	0	0	0	0	0	0	0	0	0	0	0	0	0	2
OCT	0	2	1	0	0	0	0	0	0	0	0	0	0	0	0	0	0	0	0	1	0	0
NOV	0	0	0	0	0	0	0	1	0	0	0	0	0	0	0	0	0	0	0	0	0	0
DEC	0	2	0	0	0	0	0	0	0	0	0	0	0	0	0	0	0	0	0	2	1	0
TOTAL	1	9	2	1	0	0	0	1	0	0	4	0	0	0	0	1	7	1	1	6	6	5

1987	AFL	AFL	AFL	AFL	AFL/GA	GA	SC	SC/NC	NC	NC	NC/VA	VA	MD/DE	NJ	NJ/NY	T/RI/MA	MA/NH	ME		P.RICO
ZONE	26	27	28	29	30	31	32	33	34	35	36	37	38	39	40	41	42	43	TOTAL	99
MONTH																				
JAN	1	0	0	0	0	0	0	0	0	0	0	0	0	0	0	0	0	0	6	1
FEB	3	0	0	0	0	0	0	0	0	0	0	0	0	0	0	0	0	0	7	1
MAR	1	0	1	0	0	0	0	0	0	0	0	0	0	0	0	0	0	0	7	0
APR	2	4	0	0	0	0	0	0	0	0	0	0	0	0	0	0	0	0	13	1
MAY	2	7	1	0	1	0	0	0	0	0	0	0	0	0	0	0	0	0	16	2
JUN	2	2	0	0	0	0	0	0	0	0	0	0	0	0	0	0	0	0	6	1
JUL	3	2	3	0	1	0	0	0	0	1	0	0	0	0	0	0	0	0	14	2
AUG	3	3	0	0	1	0	0	0	0	0	1	0	0	0	0	0	0	0	9	1
SEP	1	4	0	0	0	0	0	0	0	0	0	0	0	0	0	0	0	0	7	0
OCT	2	2	0	0	0	0	0	0	0	0	0	0	0	0	0	0	0	0	8	1
NOV	1	0	0	0	0	0	0	0	0	0	0	0	0	0	0	0	0	0	2	0
DEC	3	1	0	0	1	0	0	0	0	0	0	0	0	0	0	0	0	0	10	0
TOTAL	24	25	5	0	4	0	0	0	0	1	1	0	0	0	0	0	0	0	105	10

LEATHERBACK

1987	TX	TX	TX	TX	LA	LA	LA	LA	LA	LA/MS	MS/AL	AL/GFL	GFL	GFL	GFL	GFL	GFL	GFL	GFL	KEYS	KEYS	AFL
ZONE	21	20	19	18	17	16	15	14	13	12	11	10	9	8	7	6	5	4	3	1	24	25
MONTH																						
JAN	0	0	0	0	0	0	0	0	0	0	0	0	0	0	0	0	0	0	0	0	0	0
FEB	0	0	0	0	0	0	0	0	0	0	0	0	0	0	0	0	0	0	0	0	0	0
MAR	0	0	0	0	0	0	0	0	0	0	0	0	0	0	0	0	0	0	0	0	0	0
APR	0	0	0	0	0	0	0	0	0	0	0	0	1	0	0	0	0	0	0	0	0	0
MAY	0	0	0	0	1	0	0	0	0	0	0	0	0	0	0	0	0	0	0	0	0	0
JUN	0	0	0	0	0	0	0	0	0	0	0	0	0	0	0	0	0	0	0	0	0	0
JUL	0	0	0	0	0	0	0	0	0	0	0	0	0	0	0	0	0	0	0	0	0	0
AUG	0	0	0	0	0	0	0	0	0	0	0	0	0	0	0	0	0	0	0	0	0	0
SEP	0	0	0	0	0	0	0	0	0	0	0	0	0	0	0	0	0	0	0	0	0	0
OCT	0	0	0	0	0	0	0	0	0	0	0	0	0	0	0	0	0	0	0	0	0	0
NOV	0	0	0	1	0	0	0	0	0	0	0	0	0	0	0	0	0	0	0	0	0	0
DEC	0	0	0	0	0	0	0	0	0	0	0	0	0	0	0	0	0	0	0	0	0	0
TOTAL	0	0	0	1	1	0	0	0	0	0	0	0	1	0	0	0	0	0	0	0	0	0

1987	AFL	AFL	AFL	AFL	AFL/GA	GA	SC	SC/NC	NC	NC	NC/VA	VA	MD/DE	NJ	NJ/NY	T/RI/MA	MA/NH	ME		P.RICO
ZONE	26	27	28	29	30	31	32	33	34	35	36	37	38	39	40	41	42	43	TOTAL	99
MONTH																				
JAN	0	0	0	0	0	0	0	0	0	0	0	0	0	0	0	0	0	0	0	0
FEB	0	0	0	0	0	0	0	0	0	0	0	0	0	0	0	0	0	0	0	0
MAR	0	0	0	0	0	0	0	0	0	0	0	0	0	0	0	0	0	0	0	0
APR	0	0	0	0	11	0	0	1	0	0	0	0	0	0	0	0	0	0	13	1
MAY	0	0	0	0	1	16	2	0	0	0	0	0	0	0	0	0	0	0	20	0
JUN	0	0	0	0	0	2	0	0	0	0	1	0	0	0	1	0	0	0	4	0
JUL	0	0	0	0	0	0	0	0	0	0	0	1	1	2	3	2	0	0	9	0
AUG	0	0	0	0	0	0	0	0	0	0	0	0	1	1	12	3	0	0	17	0
SEP	0	0	0	0	0	0	0	0	0	0	0	0	1	4	6	0	0	0	11	0
OCT	0	0	0	1	0	0	0	0	0	0	0	0	1	6	5	0	1	0	14	0
NOV	0	0	0	9	8	0	0	0	0	0	2	0	0	1	1	0	0	0	22	0
DEC	0	0	0	6	3	0	0	0	0	0	0	0	0	0	0	0	0	0	9	0
TOTAL	0	0	0	16	23	18	2	1	0	0	3	1	4	14	28	5	1	0	119	1

HAWKSBILL

1987	TX	TX	TX	TX	LA	LA	LA	LA	LA	LA/MS	MS/AL	AL/GFL	GFL	GFL	GFL	GFL	GFL	GFL	GFL	KEYS	KEYS	AFL
ZONE	21	20	19	18	17	16	15	14	13	12	11	10	9	8	7	6	5	4	3	1	24	25
MONTH																						
JAN	0	0	0	0	0	0	0	0	0	0	0	0	0	0	0	0	0	0	0	0	0	0
FEB	0	0	1	0	0	0	0	0	0	0	0	0	0	0	0	0	0	0	0	0	0	0
MAR	0	0	0	0	0	0	0	0	0	0	0	0	0	0	0	0	0	0	0	0	0	1
APR	0	0	0	0	0	0	0	0	0	0	0	0	0	0	0	0	0	0	0	0	0	0
MAY	0	0	0	0	0	0	0	0	0	0	0	0	0	0	0	0	0	0	0	0	0	0
JUN	1	0	0	0	0	0	0	0	0	0	0	0	0	0	0	0	0	0	0	0	0	0
JUL	0	2	1	0	0	0	0	0	0	0	0	0	0	0	0	0	0	0	0	0	0	0
AUG	0	2	0	0	0	0	0	0	0	0	0	0	0	0	0	0	0	0	0	0	0	1
SEP	0	0	0	0	0	0	0	1	0	0	0	0	0	0	0	0	0	0	0	0	0	0
OCT	1	1	0	0	0	0	0	0	0	0	0	0	0	0	0	0	0	0	0	0	0	0
NOV	0	0	0	0	0	0	0	0	0	0	0	0	0	0	0	0	0	0	0	0	0	0
DEC	0	1	0	0	0	0	0	0	0	0	0	0	0	0	0	0	0	0	0	0	0	0
TOTAL	2	6	2	0	0	0	0	1	0	0	0	0	0	0	0	0	0	0	0	0	0	2

1987	AFL	AFL	AFL	AFL	AFL/GA	GA	SC	SC/NC	NC	NC	NC/VA	VA	MD/DE	NJ	NJ/NY	T/RI/MA	MA/NH	ME		P.RICO
ZONE	26	27	28	29	30	31	32	33	34	35	36	37	38	39	40	41	42	43	TOTAL	99
MONTH																				
JAN	0	0	0	0	0	0	0	0	0	0	0	0	0	0	0	0	0	0	0	0
FEB	0	0	0	0	0	0	0	0	0	0	0	0	0	0	0	0	0	0	1	0
MAR	1	0	0	0	0	0	0	0	0	0	0	0	0	0	0	0	0	0	2	0
APR	0	0	1	0	0	0	0	0	0	0	0	0	0	0	0	0	0	0	1	0
MAY	0	0	0	0	1	0	0	0	0	0	0	0	0	0	0	0	0	0	1	0
JUN	0	0	0	0	0	0	0	0	0	0	0	0	0	0	0	0	0	0	1	0
JUL	0	0	0	0	0	0	0	0	0	0	0	0	0	0	0	0	0	0	3	0
AUG	1	0	0	0	0	0	0	0	0	0	0	0	0	0	0	0	0	0	4	3
SEP	1	0	1	0	0	1	0	0	0	0	0	0	0	0	0	0	0	0	4	1
OCT	0	0	0	0	0	0	0	0	0	0	0	0	0	0	0	0	0	0	2	0
NOV	0	0	1	1	0	0	0	0	0	0	0	0	0	0	0	0	0	0	2	0
DEC	0	0	0	0	0	0	0	0	0	0	0	0	0	0	0	0	0	0	1	0
TOTAL	3	0	3	1	1	1	0	0	0	0	0	0	0	0	0	0	0	0	22	4

SEA TURTLES

1987	TX	TX	TX	TX	LA	LA	LA	LA	LA	LA/MS	MS/AL	AL/GFL	GFL	GFL	GFL	GFL	GFL	GFL	GFL	KEYS	KEYS	AFL
ZONE	21	20	19	18	17	16	15	14	13	12	11	10	9	8	7	6	5	4	3	1	24	25
MONTH																						
JAN	0	0	0	0	1	0	0	0	0	0	0	0	0	0	0	1	3	2	0	0	1	0
FEB	3	4	2	0	0	0	0	0	0	0	0	0	0	0	0	0	1	5	1	1	1	0
MAR	5	5	1	1	2	0	0	0	0	0	0	0	0	0	0	0	12	20	0	1	2	2
APR	13	36	12	3	2	0	0	0	0	0	3	1	1	1	1	0	12	11	0	0	0	0
MAY	5	14	8	6	11	0	0	2	2	0	12	11	0	4	2	0	8	11	0	1	1	3
JUN	1	4	3	4	4	0	0	4	1	0	7	24	0	2	0	0	6	5	0	0	0	1
JUL	2	12	13	4	4	0	0	0	0	0	1	0	0	1	0	0	2	4	0	2	1	1
AUG	1	6	5	7	2	0	1	0	0	0	0	0	1	0	1	0	3	2	0	1	0	1
SEP	0	3	5	5	2	0	0	1	2	0	0	0	0	1	0	0	0	1	0	0	1	2
OCT	2	5	7	4	0	2	0	0	0	0	7	0	0	0	0	0	0	2	0	1	0	0
NOV	0	2	1	3	0	3	0	1	0	0	0	0	0	0	0	0	2	0	0	0	0	0
DEC	3	8	4	2	0	0	9	1	0	0	0	0	0	0	0	0	0	3	0	2	1	2
TOTAL	35	99	61	39	28	5	10	9	5	0	30	36	2	9	4	1	49	66	1	9	8	12

1987	AFL	AFL	AFL	AFL	AFL/GA	GA	SC	SC/NC	NC	NC	NC/VA	VA	MD/DE	NJ	NJ/NY	T/RI/MA	MA/NH	ME		P.RICO
ZONE	26	27	28	29	30	31	32	33	34	35	36	37	38	39	40	41	42	43	TOTAL	99
MONTH																				
JAN	2	0	1	1	1	0	0	0	0	11	0	0	0	0	0	0	0	0	24	1
FEB	8	1	6	1	0	0	0	0	0	2	0	0	0	0	0	0	0	0	36	1
MAR	4	3	5	2	0	0	0	0	5	2	0	0	0	0	0	0	0	0	72	0
APR	4	9	3	5	18	1	0	1	0	0	0	0	0	0	0	0	0	0	137	2
MAY	5	20	8	6	62	107	22	24	12	1	2	0	0	0	0	0	0	0	370	2
JUN	2	8	11	44	54	99	131	13	22	2	16	3	0	0	3	0	0	0	474	10
JUL	4	6	12	33	42	25	44	24	13	1	3	1	2	5	6	2	0	0	270	2
AUG	5	9	7	8	40	12	11	11	13	3	4	5	4	7	17	3	0	0	190	4
SEP	2	6	5	11	30	8	14	15	3	0	4	0	6	8	11	1	0	0	147	1
OCT	3	3	0	4	7	8	3	1	0	3	8	0	4	9	7	0	1	0	91	1
NOV	2	2	3	26	36	6	0	1	1	1	4	0	0	1	2	0	0	0	97	0
DEC	4	6	0	18	15	1	0	0	2	15	0	0	0	0	0	0	0	0	96	0
TOTAL	45	73	61	159	305	267	225	90	71	41	41	9	16	30	46	6	1	0	2004	24

LOGGERHEAD

NOTES: 1988 STRANDINGS OFFSHORE WITHOUT HEADSTART TURTLES, unidentified included in total sea turtles.

1988	TX	TX	TX	TX	LA	LA	LA	LA	LA	LA/MS	MS/AL	AL/GFL	GFL	GFL	GFL	GFL	GFL	GFL	GFL	KEYS	KEYS	AFL	AFL
ZONE	21	20	19	18	17	16	15	14	13	12	11	10	9	8	7	6	5	4	3	1	24	25	26
MONTH																							
JAN	1	0	2	0	0	0	0	1	0	0	0	0	0	0	0	0	0	0	0	0	1	0	4
FEB	1	1	0	0	0	0	0	1	0	0	0	0	0	0	0	0	0	1	0	0	0	0	2
MAR	2	7	4	0	0	0	0	1	0	0	0	0	0	1	0	1	1	2	0	0	2	0	3
APR	8	27	4	0	0	0	0	0	0	0	2	0	0	0	0	2	12	4	0	0	1	0	2
MAY	2	15	6	1	0	0	0	0	0	0	1	0	0	0	1	1	17	6	0	0	1	0	3
JUN	1	3	1	0	0	0	0	1	0	0	8	0	0	4	0	0	2	1	0	0	0	0	2
JUL	4	3	0	0	0	0	0	0	0	1	2	2	3	8	0	0	11	1	0	0	0	0	2
AUG	0	6	0	0	0	0	0	0	0	0	0	0	2	4	0	1	3	2	0	0	2	1	2
SEP	0	1	0	1	0	0	0	0	0	0	2	0	0	0	0	0	2	1	0	1	0	1	1
OCT	1	0	1	2	1	0	0	0	0	0	1	0	0	2	0	0	2	1	0	0	0	0	0
NOV	0	0	4	2	0	0	0	0	0	0	1	0	0	2	0	0	3	2	0	0	0	0	1
DEC	0	4	0	1	0	0	0	0	0	0	1	0	0	0	0	1	1	1	0	0	0	0	0
TOTAL	20	67	22	7	1	0	0	4	0	1	18	2	5	21	1	6	54	22	0	1	7	2	22

1988	AFL	AFL	AFL	AFL/GA	GA	SC	SC/NC	NC	NC	NC/VA	VA	MD/DE	NJ	NJ/NY	T/RI/MA	MA/NH	ME		P.RICO
ZONE	27	28	29	30	31	32	33	34	35	36	37	38	39	40	41	42	43	TOTAL	99
MONTH																			
JAN	2	5	5	1	0	0	0	1	2	0	0	0	0	0	0	0	0	25	0
FEB	9	5	1	2	0	0	0	0	0	0	0	0	0	0	0	0	0	23	0
MAR	8	15	11	0	0	0	0	0	0	0	0	0	0	0	0	0	0	58	0
APR	2	6	11	14	0	1	4	0	1	0	0	0	0	0	0	0	0	101	0
MAY	16	10	4	21	19	2	24	9	4	10	0	0	0	0	0	0	0	173	0
JUN	4	8	7	26	19	1	3	13	5	21	2	1	0	0	0	0	0	133	0
JUL	2	7	10	27	13	32	11	5	3	5	0	0	1	1	0	0	0	154	0
AUG	13	16	19	46	16	27	7	0	1	2	0	0	5	4	0	0	0	179	0
SEP	7	3	25	19	17	6	5	1	0	4	3	0	3	7	0	0	0	110	0
OCT	3	0	22	5	8	2	2	7	3	6	0	1	1	0	0	0	0	71	0
NOV	0	1	16	15	4	1	0	7	7	5	2	0	0	0	0	0	0	73	0
DEC	3	4	15	7	1	0	0	3	8	0	0	0	0	0	0	0	0	50	0
TOTAL	69	80	146	183	97	72	56	46	34	53	7	2	10	12	0	0	0	1150	0

KEMP'S RIDLEY

1988	TX	TX	TX	TX	LA	LA	LA	LA	LA	LA/MS	MS/AL	AL/GFL	GFL	GFL	GFL	GFL	GFL	GFL	GFL	KEYS	KEYS	AFL	AFL
ZONE	21	20	19	18	17	16	15	14	13	12	11	10	9	8	7	6	5	4	3	1	24	25	26
MONTH																							
JAN	0	0	0	0	0	0	0	0	0	0	0	0	0	0	0	0	0	0	0	0	0	0	0
FEB	0	1	1	0	0	0	0	0	0	0	0	0	0	0	0	0	0	0	0	0	0	0	0
MAR	0	1	0	0	0	0	0	2	0	0	0	0	0	0	0	0	0	0	0	0	0	0	0
APR	0	0	6	3	0	0	1	0	0	0	0	0	0	0	0	0	0	0	0	0	0	0	0
MAY	1	6	3	0	1	0	1	0	0	0	0	0	0	0	0	0	1	1	0	0	0	0	0
JUN	0	0	0	0	0	0	0	1	1	0	0	0	0	1	0	0	0	0	0	0	0	0	0
JUL	1	0	1	1	0	0	1	0	0	0	2	0	0	0	0	0	1	0	0	0	0	0	0
AUG	0	1	0	0	0	0	0	0	0	0	0	0	0	0	0	0	0	0	0	0	0	0	0
SEP	0	1	0	1	0	0	0	0	0	0	1	0	0	0	0	0	0	0	0	0	0	0	0
OCT	0	1	0	0	0	0	0	0	0	0	0	0	0	0	0	0	0	0	0	0	0	0	0
NOV	1	1	2	4	0	0	0	0	0	0	2	0	0	0	0	0	0	0	0	0	0	0	0
DEC	0	1	0	1	0	0	0	0	0	0	1	0	0	1	0	0	0	0	0	0	0	0	0
TOTAL	3	13	13	10	1	0	3	3	1	0	6	0	0	2	0	0	2	1	0	0	0	0	0

1988	AFL	AFL	AFL	AFL/GA	GA	SC	SC/NC	NC	NC	NC/VA	VA	MD/DE	NJ	NJ/NY	T/RI/MA	MA/NH	ME		P.RICO
ZONE	27	28	29	30	31	32	33	34	35	36	37	38	39	40	41	42	43	TOTAL	99
MONTH																			
JAN	0	0	0	1	0	0	0	0	0	0	0	0	0	0	1	0	0	2	0
FEB	0	0	0	0	0	0	0	0	0	0	0	0	0	0	0	0	0	2	0
MAR	0	1	4	0	0	0	0	0	1	0	0	0	0	0	0	0	0	9	0
APR	0	0	0	1	0	1	0	0	0	0	0	0	0	0	0	0	0	12	0
MAY	0	0	0	0	0	1	0	0	0	0	0	0	0	0	0	0	0	15	0
JUN	0	0	0	0	6	0	0	0	0	0	0	0	0	0	0	0	0	9	0
JUL	0	0	3	0	1	0	0	1	0	0	0	0	0	0	0	0	0	12	0
AUG	0	1	1	0	2	1	0	0	0	0	0	0	0	0	1	0	0	7	0
SEP	1	0	0	1	1	0	1	0	0	1	0	0	0	0	0	0	0	8	0
OCT	0	0	1	8	4	1	0	0	0	7	0	0	0	0	0	0	0	22	0
NOV	0	0	0	32	2	0	1	1	1	0	1	0	0	0	0	0	0	48	0
DEC	1	0	11	12	1	0	0	0	1	0	0	0	0	0	0	0	0	30	0
TOTAL	2	2	20	55	17	4	2	2	3	8	1	0	0	0	2	0	0	176	0

GREEN 1988	TX	TX	TX	TX	LA	LA	LA	LA	LA	LA/MS	MS/AL	AL/GFL	GFL	GFL	GFL	GFL	GFL	GFL	GFL	KEYS	KEYS	AFL	AFL
ZONE	21	20	19	18	17	16	15	14	13	12	11	10	9	8	7	6	5	4	3	1	24	25	26
MONTH																							
JAN	0	0	0	0	0	0	0	0	0	0	0	0	0	0	0	0	0	0	0	0	0	0	3
FEB	0	0	0	0	0	0	0	1	0	0	0	0	0	0	0	0	0	0	0	0	0	0	0
MAR	0	1	0	0	0	0	0	0	0	0	0	0	0	0	0	1	2	2	0	1	0	0	4
APR	0	2	0	0	0	0	0	0	0	0	0	0	0	0	0	0	0	0	1	0	0	0	1
MAY	0	1	0	0	0	0	0	0	0	0	0	1	0	0	0	0	2	0	0	0	0	0	2
JUN	0	1	1	0	0	0	0	0	0	0	1	0	0	0	0	0	0	0	0	0	1	0	2
JUL	0	0	0	0	0	0	0	0	0	0	0	0	0	0	0	0	0	0	0	1	0	0	3
AUG	0	0	0	0	0	0	0	0	0	0	0	0	0	0	0	0	0	0	0	2	0	0	4
SEP	0	0	0	0	0	0	0	0	0	0	0	1	0	0	0	0	0	0	0	0	0	0	2
OCT	0	0	0	0	0	0	0	0	0	0	0	0	0	0	0	0	0	0	0	2	0	0	2
NOV	0	0	0	0	0	0	0	0	0	0	0	0	0	0	0	0	0	1	0	0	0	1	2
DEC	0	1	0	0	0	0	0	0	0	0	0	0	0	0	0	1	0	0	0	0	1	0	2
TOTAL	0	6	1	0	0	0	0	1	0	0	1	2	0	0	0	2	4	3	1	6	2	1	27

1988	AFL	AFL	AFL	AFL/GA	GA	SC	SC/NC	NC	NC	NC/VA	VA	MD/DE	NJ	NJ/NY	T/RI/MA	MA/NH	ME		P.RICO
ZONE	27	28	29	30	31	32	33	34	35	36	37	38	39	40	41	42	43	TOTAL	99
MONTH																			
JAN	2	0	2	0	1	0	0	0	0	0	0	0	0	0	0	0	0	8	1
FEB	10	0	0	0	0	0	0	0	0	0	0	0	0	0	0	0	0	11	0
MAR	13	1	0	0	0	0	0	0	0	0	0	0	0	0	0	0	0	25	0
APR	4	0	0	4	0	0	0	0	0	0	0	0	0	0	0	0	0	12	1
MAY	21	0	0	0	1	0	0	0	2	0	0	0	0	0	0	0	0	30	0
JUN	2	0	0	0	0	0	0	1	0	0	0	0	0	0	0	0	0	9	0
JUL	6	0	0	0	0	0	0	0	0	0	0	0	0	0	0	0	0	10	0
AUG	3	1	0	0	0	0	0	0	0	0	0	0	0	0	0	0	0	10	1
SEP	2	0	1	0	1	0	0	0	0	0	0	0	0	0	0	0	0	7	0
OCT	0	0	0	0	0	0	0	2	0	0	0	0	0	0	0	0	0	6	1
NOV	4	0	1	2	0	0	0	2	0	1	0	0	0	0	0	0	0	14	0
DEC	1	0	0	0	0	0	0	1	1	0	0	0	0	0	0	0	0	8	0
TOTAL	68	2	4	6	3	0	0	6	3	1	0	0	0	0	0	0	0	150	4

LEATHERBACK

1988	TX	TX	TX	TX	LA	LA	LA	LA	LA	LA/MS	MS/AL	AL/GFL	GFL	GFL	GFL	GFL	GFL	GFL	GFL	KEYS	KEYS	AFL	AFL
ZONE	21	20	19	18	17	16	15	14	13	12	11	10	9	8	7	6	5	4	3	1	24	25	26
MONTH																							
JAN	0	0	0	0	0	0	0	0	0	0	0	0	0	0	0	0	0	0	0	0	0	0	0
FEB	0	0	0	0	0	0	0	0	0	0	0	0	0	0	0	0	0	0	0	0	0	0	0
MAR	0	0	0	0	0	0	0	0	0	0	0	0	1	0	0	0	0	0	0	0	0	0	0
APR	0	0	0	1	0	0	0	0	0	0	0	0	0	0	0	0	0	0	0	0	0	0	0
MAY	0	1	1	0	0	0	0	0	0	0	0	0	0	0	0	0	0	0	0	0	0	0	1
JUN	0	0	0	0	0	0	0	0	0	0	0	1	0	0	0	0	0	0	0	0	0	0	0
JUL	0	0	0	0	0	0	0	0	0	0	0	0	0	0	0	0	0	0	0	0	0	0	0
AUG	0	0	0	0	0	0	0	1	0	0	0	0	0	0	0	0	0	0	0	0	0	0	0
SEP	0	0	0	0	0	0	0	0	0	0	0	0	0	0	0	0	0	0	0	0	0	0	1
OCT	0	0	0	2	0	0	0	0	0	0	0	0	0	0	0	0	0	0	0	0	0	0	0
NOV	0	0	1	1	0	0	0	0	0	0	0	0	0	0	0	0	0	0	0	0	0	0	0
DEC	0	0	0	0	0	0	0	0	0	0	0	0	0	0	0	0	0	0	0	0	0	0	0
TOTAL	0	1	2	4	0	0	0	1	0	0	0	1	1	0	0	0	0	0	0	0	0	0	2

1988	AFL	AFL	AFL	AFL/GA	GA	SC	SC/NC	NC	NC	NC/VA	VA	MD/DE	NJ	NJ/NY	T/RI/MA	MA/NH	ME		P.RICO
ZONE	27	28	29	30	31	32	33	34	35	36	37	38	39	40	41	42	43	TOTAL	99
MONTH																			
JAN	0	0	0	1	0	0	0	0	0	0	0	0	0	0	0	0	0	1	0
FEB	1	0	1	1	0	0	0	0	0	0	0	0	0	0	0	0	0	3	0
MAR	0	1	1	0	0	0	0	0	0	0	0	0	0	0	0	0	0	3	1
APR	0	0	0	0	0	0	0	0	0	0	0	0	0	0	0	0	0	1	0
MAY	0	0	0	0	0	1	0	0	0	0	0	0	1	2	0	0	0	7	0
JUN	0	0	1	0	1	0	0	0	0	1	0	0	1	0	0	0	0	5	0
JUL	0	0	0	0	0	0	0	0	0	1	0	0	0	1	0	0	0	2	0
AUG	0	0	0	0	0	0	0	0	0	1	0	1	0	4	2	0	0	9	0
SEP	1	0	0	0	0	0	0	0	0	0	0	0	1	3	0	0	0	6	0
OCT	0	0	8	1	0	0	0	0	0	0	0	0	0	1	1	0	0	13	0
NOV	0	0	3	3	0	0	0	0	0	0	0	0	0	2	0	0	0	10	0
DEC	0	0	1	0	0	0	0	0	0	0	0	0	0	1	1	0	0	3	0
TOTAL	2	1	15	6	1	1	0	0	0	3	0	1	3	14	4	0	0	63	1

HAWKSBILL 1988	TX	TX	TX	TX	LA	LA	LA	LA	LA	LA/MS	MS/AL	AL/GFL	GFL	GFL	GFL	GFL	GFL	GFL	GFL	KEYS	KEYS	AFL	AFL
ZONE	21	20	19	18	17	16	15	14	13	12	11	10	9	8	7	6	5	4	3	1	24	25	26
MONTH																							
JAN	0	0	0	0	0	0	0	0	0	0	0	0	0	0	0	0	0	0	0	0	0	0	0
FEB	0	0	0	0	0	0	0	0	0	0	0	0	0	0	0	0	0	0	0	0	0	0	1
MAR	0	0	0	0	0	0	0	0	0	0	0	0	0	0	0	0	0	0	0	0	0	0	0
APR	0	0	0	0	0	0	0	0	0	0	0	0	0	0	0	0	0	0	0	0	0	0	0
MAY	0	1	0	0	0	0	0	0	0	0	0	0	0	0	0	0	0	1	0	0	0	0	0
JUN	0	2	0	0	0	0	0	0	0	0	0	0	0	0	0	0	0	0	0	0	0	0	2
JUL	0	1	0	0	0	0	0	0	0	0	0	0	0	0	0	0	0	0	0	0	0	1	0
AUG	0	0	0	0	0	0	0	0	0	0	0	0	0	0	0	0	0	0	0	0	0	0	0
SEP	0	3	0	0	0	0	0	0	0	0	0	0	0	0	0	0	0	0	0	0	0	0	0
OCT	0	1	0	0	0	0	0	0	0	0	0	0	0	0	0	0	0	0	0	0	0	0	0
NOV	0	1	0	0	0	0	0	0	0	0	0	0	0	0	0	0	0	0	0	0	0	0	0
DEC	1	2	0	0	0	0	0	0	0	0	0	0	0	0	0	0	0	0	0	0	1	0	0
TOTAL	1	11	0	0	0	0	0	0	0	0	0	0	0	0	0	0	0	1	0	0	1	1	3

1988	AFL	AFL	AFL	AFL/GA	GA	SC	SC/NC	NC	NC	NC/VA	VA	MD/DE	NJ	NJ/NY	T/RI/MA	MA/NH	ME		P.RICO
ZONE	27	28	29	30	31	32	33	34	35	36	37	38	39	40	41	42	43	TOTAL	99
MONTH																			
JAN	0	0	0	0	0	0	0	0	0	0	0	0	0	0	0	0	0	0	0
FEB	0	0	0	0	0	0	0	0	0	0	0	0	0	0	0	0	0	1	0
MAR	0	0	0	0	0	0	0	0	0	0	0	0	0	0	0	0	0	0	1
APR	1	0	0	0	0	0	0	0	0	0	0	0	0	0	0	0	0	1	0
MAY	0	0	0	0	0	0	1	0	0	0	0	0	0	0	0	0	0	3	1
JUN	0	0	0	0	0	0	0	0	0	0	0	0	0	0	0	0	0	4	0
JUL	0	0	0	0	0	0	0	0	0	0	0	0	0	0	0	0	0	2	0
AUG	0	0	0	0	0	0	0	0	0	0	0	0	0	0	0	0	0	0	0
SEP	0	0	0	0	0	0	0	0	0	0	0	0	0	0	0	0	0	3	1
OCT	0	0	0	0	0	0	0	0	0	0	0	0	0	0	0	0	0	1	0
NOV	0	0	0	0	0	0	0	0	0	0	0	0	0	0	0	0	0	1	0
DEC	0	0	0	0	0	0	0	0	0	0	0	0	0	0	0	0	0	4	1
TOTAL	1	0	0	0	0	0	1	0	0	0	0	0	0	0	0	0	0	20	4

SEA TURTLES

1988	TX	TX	TX	TX	LA	LA	LA	LA	LA	LA/MS	MS/AL	AL/GFL	GFL	GFL	GFL	GFL	GFL	GFL	GFL	KEYS	KEYS	AFL	AFL
ZONE	21	20	19	18	17	16	15	14	13	12	11	10	9	8	7	6	5	4	3	1	24	25	26
MONTH																							
JAN	1	0	2	0	0	0	2	1	0	0	0	0	0	0	0	0	0	2	0	0	1	0	7
FEB	1	2	1	0	0	0	0	2	0	0	0	0	0	0	0	0	0	1	0	0	0	0	4
MAR	2	9	5	0	0	0	0	3	0	0	0	0	1	1	0	2	3	4	0	1	2	0	7
APR	8	31	10	4	0	0	1	0	0	0	2	0	0	0	0	2	12	4	1	0	1	0	3
MAY	3	24	10	1	1	0	1	3	0	0	1	1	0	0	1	1	20	8	0	0	1	0	6
JUN	1	8	2	0	0	0	0	2	1	1	11	1	0	5	0	0	2	1	0	0	1	0	6
JUL	5	4	1	1	0	0	1	0	0	1	8	2	3	8	0	0	12	1	0	1	0	1	5
AUG	0	7	1	0	0	0	0	1	0	0	0	0	2	4	0	1	3	2	0	2	2	1	6
SEP	0	5	0	2	0	0	0	0	0	0	4	1	0	0	0	0	2	1	0	1	0	1	4
OCT	1	3	1	4	1	0	0	0	0	0	1	0	0	2	0	0	2	1	0	2	0	0	2
NOV	1	2	7	7	0	0	0	0	0	0	3	0	0	2	0	0	3	3	0	0	0	1	3
DEC	1	8	0	2	0	0	0	0	0	0	2	0	0	1	0	2	1	1	0	0	2	0	2
TOTAL	24	103	40	21	2	0	5	12	1	2	32	5	6	23	1	8	60	29	1	7	10	4	55

1988	AFL	AFL	AFL	AFL/GA	GA	SC	SC/NC	NC	NC	NC/VA	VA	MD/DE	NJ	NJ/NY	T/RI/MA	MA/NH	ME		P.RICO
ZONE	27	28	29	30	31	32	33	34	35	36	37	38	39	40	41	42	43	TOTAL	99
MONTH																			
JAN	4	5	7	3	1	0	0	1	2	0	0	0	0	0	1	0	0	40	1
FEB	20	5	2	3	0	0	0	0	0	0	0	0	0	0	0	0	0	41	0
MAR	21	19	16	0	0	0	0	0	1	0	0	0	0	0	0	0	0	97	2
APR	7	7	12	19	0	2	4	0	1	0	0	0	0	0	0	0	0	131	1
MAY	40	13	5	22	20	4	26	9	6	10	0	0	1	2	0	0	0	240	1
JUN	6	8	8	26	26	2	3	14	5	23	2	1	1	0	0	0	0	167	0
JUL	8	7	14	27	15	34	12	7	3	6	0	0	1	2	0	0	0	190	0
AUG	16	18	21	48	19	29	7	0	1	3	0	1	5	8	3	0	0	211	1
SEP	11	3	27	23	22	6	6	2	0	5	3	0	4	10	0	0	0	143	1
OCT	3	0	36	15	12	3	2	9	3	13	0	1	1	1	1	0	0	120	1
NOV	4	1	21	54	6	1	1	10	8	6	3	0	0	2	0	0	0	149	0
DEC	5	4	30	19	2	0	0	4	10	0	0	0	0	1	1	0	0	98	1
TOTAL	145	90	199	259	123	81	61	56	40	66	8	3	13	26	6	0	0	1627	9

Appendix F

Shrimp Fishing Effort, 1987-1988

Compiled by the committee from unpublished raw NMFS data.

NOTES: 1988 SHRIMP FISHING EFFORT 24 HOURS OF TOWING NETS DOES NOT INCLUDE BAYS

1988	TX	TX	TX	TX	LA	LA	LA	LA	LA	LA/MS	MS/AL	AL/GFL	GFL	GFL	GFL	GFL	GFL	GFL	GFL	OFSR	KEYS	KEYS
ZONE	21	20	19	18	17	16	15	14	13	12	11	10	9	8	7	6	5	4	3	2	1	24
MONTH																						
JAN	433	346	896	503	325	1094	624	490	889	44	449	0	0	27	123	67	16	177	394	1104	28	
FEB	775	671	1292	433	495	760	952	600	469	6	539	19	0	18	174	49	8	179	708	1075	11	
MAR	555	325	578	235	357	686	466	250	278	20	528	8	19	45	87	22	8	277	534	1398	16	
APR	728	508	820	394	667	666	515	287	610	4	1020	13	10	54	263	82	179	248	363	1491	51	
MAY	1447	1688	1580	749	994	1382	1919	2005	5507	556	958	26	19	148	385	93	248	255	391	1431	45	
JUN	1219	1029	1522	645	1978	4334	2268	1522	2166	658	1498	178	0	49	534	234	181	185	218	745	102	
JUL	2088	2348	3965	2681	1416	2139	1124	1713	2385	288	1080	43	0	217	341	63	72	26	42	244	17	
AUG	2110	1984	3653	1933	3335	3865	2097	2277	2270	134	2012	50	0	110	240	64	48	26	63	299	2	
SEP	1797	1442	2048	1898	2515	3715	1635	1235	933	330	1661	0	0	68	176	95	45	24	62	241	4	
OCT	2562	1391	2295	2710	3997	4868	3329	1373	2774	324	1186	31	23	85	368	120	96	40	117	396	5	
NOV	1287	967	2409	1527	2884	3278	2404	799	1929	221	1325	86	0	271	346	80	121	124	101	788	0	
DEC	1357	1063	2020	678	875	2351	2296	853	1326	88	1062	5	8	109	313	100	70	341	343	818	60	
TOTAL	15001	13762	23078	14386	19838	29138	19629	13404	21536	2673	13318	459	79	1201	3350	1069	1092	1902	3336	10030	341	

1988	AFL	AFL	AFL	AFL	AFL	AFL/GA	GA	SC	SC/NC	NC	NC	NC/VA	VA	MD/DE	NJ	NJ/NY	T/RI/MA	MA/NH	ME	
ZONE	25	26	27	28	29	30	31	32	33	34	35	36	37	38	39	40	41	42	43	TOTAL
MONTH																				
JAN				123	42	238	0	142	4	0	0									8577.3
FEB				125	43	91	0	0	0	0	0									9492.7
MAR				187	65	37	0	0	0	0	4									6985.2
APR				18	30	63	2	0	0	43	34									9162.6
MAY				16	18	229	5	17	0	28	69									22207
JUN				60	61	303	2	104	10	42	280									22128
JUL				18	163	343	0	821	188	48	98									23971
AUG				27	106	375	623	855	246	97	61									28961
SEP				25	111	394	482	1101	122	47	69									22274
OCT				64	233	291	875	1040	49	42	52									30736
NOV				140	144	463	739	725	20	27	43									23248
DEC				188	250	428	895	752	20	7	9									18685
TOTAL				989	1266	3256	3623	5558	658	380	719									226427

NOTES: 1987 SHRIMP FISHING EFFORT 24 HOURS OF TOWING NETS. DOES NOT INCLUDE BAYS.

1987	TX	TX	TX	TX	LA	LA	LA	LA	LA	LA/MS	MS/AL	AL/GFL	GFL	GFL	GFL	GFL	GFL	GFL	GFL	OFSA	KEYS	KEYS
ZONE	21	20	19	18	17	16	15	14	13	12	11	10	9	8	7	6	5	4	3	2	1	24
MONTH																						
JAN	744	1073	2353	494	638	672	990	763	954	41	413	9	0	96	103	27	29	281	356	1154	18	
FEB	1261	978	1112	823	698	675	1257	846	850	32	239	0	0	72	311	22	362	270	257	2150	46	
MAR	829	266	1080	635	903	1408	700	692	1002	29	402	12	0	190	266	128	229	484	374	1474	23	
APR	823	713	1289	527	1187	851	1037	694	813	21	391	0	0	145	414	61	323	616	249	1330	21	
MAY	1037	1222	1950	871	1312	1260	2378	1611	6080	363	665	18	18	185	657	189	332	399	190	1041	11	
JUN	1208	775	3456	1043	2762	2546	6985	1160	5094	641	1166	399	0	122	255	116	67	219	197	772	5	
JUL	1876	1588	5072	3482	1678	2455	3256	1120	3365	97	1024	118	23	250	138	17	84	91	60	338	3	
AUG	2041	1705	3626	1977	3469	2739	3501	1230	2813	124	1011	79	0	158	371	44	46	38	146	457	12	
SEP	1564	1505	3189	1419	4655	2422	3422	1067	6830	221	1062	17	0	46	56	36	131	48	167	376	116	
OCT	1663	1335	2865	2366	4791	2904	3179	1937	6596	265	1112	153	0	48	150	47	25	60	147	763	106	
NOV	1379	1082	2766	2503	2620	2398	2710	1816	3472	297	704	31	0	52	93	101	9	99	239	1133	87	
DEC	1472	739	2970	1457	1074	1740	1802	1812	2236	420	785	2	33	104	81	15	7	280	691	1512	24	
TOTAL	14425	12981	31728	17597	25787	22070	31217	14748	40105	2551	8974	838	74	1468	2895	803	1644	2885	3073	12500	472	

1988	AFL	AFL	AFL	AFL	AFL	AFL/GA	GA	SC	SC/NC	NC	NC	NC/VA	VA	MD/DE	NJ	NJ/NY	T/RI/MA	MA/NH	ME	
ZONE	25	26	27	28	29	30	31	32	33	34	35	36	37	38	39	40	41	42	43	TOTAL
MONTH																				
JAN			0	199	198	325	641	297	7	0										12874
FEB			0	75	82	115	68	13	1	0										12615
MAR			0	15	32	41	19	0	0	3										11236
APR			0	9	6	48	76	2	2	72										11720
MAY			0	12	12	198	828	94	2	104										23040
JUN			0	5	52	213	914	798	100	324										31394
JUL			1	43	57	236	610	883	438	357										28760
AUG			0	50	76	230	735	704	218	182										27782
SEP			0	216	87	199	1246	1365	436	124										32022
OCT			7	128	95	265	996	1466	144	138										33751
NOV			9	257	236	346	669	857	107	70										26141
DEC			0	269	152	296	543	598	59	21										21193
TOTAL			17	1277	1085	2512	7345	7076	1513	1395										272528

Appendix G

Annotated Chronological List of Educational Efforts on TEDs for Fishermen by the National Marine Fisheries Service

1977

Joint meetings held among shrimp industry, environmental community, and NMFS representatives to find way to keep turtles from drowning that would not hurt the shrimp industry.

1980

Scoping meetings in Atlanta and Richmond Hill, Georgia-October (USDOC/NMFS 1983). Identified broad range of issues to be considered in formulation of actions designed to reduce incidental catch of turtles in commercial shrimp fishery.

Voluntary TED program.

Technology Transfer Program (Oravetz, 1983a)

September	Charleston, South Carolina-TED data presented to shrimp industry and environmental community
October	Emergency resuscitation regulations

Compiled by Nancy Balcom and William DuPaul, Virginia Institute of Marine Science, and College of William and Mary.

1981

TED workshops (pers. comm., C. Oravetz)

April 2	Brunswick, Georgia, for Sea Grant agents and GA DNR
April 7	Charleston, South Carolina, for Sea Grant agents and South Carolina Wildlife and Marine Resources Department
April 8	Manteo, North Carolina, for Sea Grant agents (Coastwatch, 1981) Technology Transfer Program (Oravetz, 1983a)
March	Through April—Sea Grant workshops for Georgia, Florida, South Carolina, North Carolina providing information packet, videotape of TEDs, fisherman contacts
August	Through October—TED demonstration contract; TEDs demonstrated to 196 fishermen in 13 ports in South Carolina, Georgia, Florida (7-8 TED users resulted)

1981-1982

100 TEDs built and distributed to fishermen in South Atlantic states.

1982

TED workshops (pers. comm., C. Oravetz, NMFS, 1983)

September 30 St. Augustine, Florida, for Sea Grant agents and MAP Coordinators Technology Transfer Program (Oravetz, 1983a)

May	TED committee formed between shrimp industry, environmental groups, Sea Grant and NMFS to assist and monitor progress
June	TED booth at Southeastern Fisheries Association convention (NMFS, 1983); TED displayed and information provided to potential users

NMFS TED construction: 96 additional TEDs distributed with webbing

NMFS TED Technology Transfer Program (pers. comm., C. Oravetz , 1982)

October	North Carolina Commercial Fisheries Show
	Three TED displays at North Carolina Marine Resource Centers
	TED on display at Cape Canaveral Marine and Equipment Supply Shop
February	Dayton—"Coastal Living" mall exhibit
April	Miami—commercial boat show
July	Costa Rica—poster session at WATS symposium
December	Helen, Georgia, SG/NMFS Retreat

1983

TED workshops (pers. comm., C. Oravetz; Oravetz, 1983a)

March	Louisiana for Sea Grant agents
April 13	Pascagoula, Mississippi for Mississippi-Alabama Sea Grant agents

TED Technology Transfer Program (pers. comm., C. Oravetz, 1982)
Direct contacts:
Personal visits to Florida fishermen Oct-Nov
Personal visits by North Carolina Marine Advisory agent
Phone and letter surveys of TED users
Personal letter responses to inquiries
Letters to net shops

Technical assistance:
TED slides to Florida Department of Natural Resources
Distribution of DESCO TEDs
Lists of TED users to Sea Grant agents

Television:
Talk show with slides (Orlando)
Taped educational TV program
Lorne Green's wilderness show

January	Workshop for Louisiana Marine Advisory agents

SAMFC

Louisiana Shrimpers Assoc. Annual Convention

Concerned Shrimpers of Louisiana asked NMFS to present a program on development and use of TEDs at annual convention

April	Workshop for Mississippi/Alabama Marine Advisory agents
	Charleston, South Carolina—sea turtle research workshop
June	Survey results of 80 TED recipients (30 responses) (Oravetz, 1983a)
July	WATS symposium in Costa Rica

Bryan Fishermen's Cooperative, Inc. (Georgia) arranged TED demonstrations aboard member vessels

NMFS and SAFDF contracted with DESCO marine to build 200 TEDs for distribution to commercial fishermen (NMFS, 1983; Oravetz, 1983b)

January	Florida: 60 Georgia: 44 North Carolina: 25 South Carolina: 42 Louisiana: 4 Texas: 2

1984

TED workshops (pers. comm., C. Oravetz)

January	Mobile, Alabama, for Sea Grant agents and fishermen
September	Pascagoula, Mississippi, demonstration for 11 Sea Grant agents, 6 net-shop owners and fishermen, and 6 NMFS personnel

TED Technology Transfer Program (pers. comm., C. Oravetz, 1982)

January	Two workshops for fishermen in Alabama

1985

TED workshops (pers. comm., C. Oravetz)

October	Charleston, South Carolina, for South Carolina Wildlife and Marine Resources Department

1986

TED workshops (pers. comm., C. Oravetz)

April	Gainesville, Florida, for Sea Grant agents

1987

Public hearings on proposed regulations requiring fishermen to use TEDs

March	Key West; Cape Canaveral; Brunswick; Morehead City; Galveston; Cameron, Louisiana; Corpus Christi; Houma, Louisiana; Kenner, Louisiana; Port Isabel, Texas; Bayou La Batre, Alabama (Wallace and Hosking, 1987a); Charleston, Washington, D.C. (Cottingham and Oravetz, 1987; Little, 1987; NMFS, 1987)
September	Compared two standard nets with one Georgia jumper and one Morrison soft TED on fisherman's boat (4 drags, 2 nights) (Little, 1987)

Series of mediation meetings held between environmental groups, shrimp associations, and NMFS (Cottingham and Oravetz, 1987; Little, 1987; NMFS, 1987)

October	New Orleans
October	Through November—Jekyll Island, Georgia
November	Washington, D.C.
December	Houston

1988

July	TED Evaluation Program continued: two trips in Louisiana, one trip in Florida (to date, 50 trips completed over years with 65 observer days in Georgia/South Carolina, 24 in Key West, 24 in E. Florida, 43 in W. Florida, 126 in Texas, and 115 in Louisiana) (NMFS, 1989)

Personal letter: About 21 NMFS employees have spent their careers working with TEDs or educating the public about sea turtle conservation; 3 full-time TED gear specialists at Pascagoula Lab employed specifically to help fishermen and net shops build and use TEDs; 4 additional gear people at Pascagoula have conducted the at-sea TED demonstrations and have given many presentations over the years; NMFS has cooperated on a regular basis with fishing industry associations, SG, state and environmental groups; NMFS relies heavily on SG for role in TED Technology Transfer - distribution, experimentation, training, information dissemination; provided 300-400 free TEDs to individual fishermen over years to test and use, have slide programs, videotapes, brochures, instructional materials available on TEDs and sea turtle conservation; conducted countless workshops and demonstrations for shrimp fishermen, SG agents, state and university personnel, reporters, commercial fisheries association meetings and conventions, on docks and on decks of vessels (pers. comm., C. Oravetz, 1989).

Appendix H

Annotated Chronological List of Educational Efforts on TEDs for Fishermen by Sea Grant

1981

Texas Sea Grant

Looking to improve TEDs—aluminum, fiberglass, collapsible—review methods to improve finfish separation (Martin, 1986)

1982

Texas Shrimp Association (USDOC/NMFS, 1983)

Supported construction and distribution of 200 TEDs

Coordinated demonstration of prototype NMFS TED.

1984

Alabama Sea Grant

January	Bayou La Batre—TED workshop (Wallace and Hosking, 1984)
	Summerdale—TED workshop
	NMFS underwater file of TED in action, slide presentation of exclusion of finfish, 2 TEDs on display (Wallace and Hosking, 1984)

Compiled by Nancy Balcom and William DuPaul, Virginia Institute of Marine Science, and College of William and Mary.

North Carolina Sea Grant

Stepped up efforts to educate fishermen (daily news item, conducted workshops at ports and shrimp houses (about 8) (pers. comm., Bahen, 1989)

1985

Alabama Sea Grant

July 6	Two TEDs to fishermen in Lafitte, LA, who had two more built (Wallace and Hosking, 1985a)
August	Four TEDs placed on Alabama shrimp boats heading for TX to test reduction of by-catch (Wallace and Hosking, 1985a)
	Six TEDs being constructed and available for local fishermen (Wallace and Hosking, 1985b)
	Looking for fishermen to test TEDs sized for 25 inch trawl against standard nets with NMFS and SG observers (Wallace and Hosking, 1985b)

Texas Sea Grant

Demonstrated NMFS TEDs on four Texas vessels

1986

Alabama Sea Grant

September	Reported on *Georgia Bulldog* cruise to test four TED types (Georgia, Matagorda, Cameron, NMFS) (Wallace and Hosking, 1986)

Georgia Sea Grant

August	Conducted comparative tests on 4 TEDs in Cape Canaveral Channel, Florida—NMFS, Georgia, Cameron, Texas TEDs (Christian, 1986; *Georgia Bulldog*, 1986)

Georgia Department of Natural Resources received $80,000 from Office of Energy Resources to purchase 221 TEDs to equip 110 shrimp boats with a pair each—87 pairs picked up by randomly selected boats (Anon., 1988a)

Texas Sea Grant

Summer — Testing 20 TED designs inshore and 10 gulf boats will test another 20 offshore (Martin, 1986)

1987

Florida Sea Grant

Holding meetings and workshops for fishermen, travelled to all major Florida port areas on Atlantic and in Gulf of Mexico to demonstrate types of TEDs, explain regulations, and advise fishermen on devices they currently use aboard their trawlers (Marine Log, 1988)

Georgia Sea Grant

March — Department of Natural Resources' second contract with Office of Energy Resources yielded $155,000, enough to provide remainder of Georgia's shrimp fleet with exception of violators (Anon., 1987). Of 248/270 boats who responded, 92% chose Georgia Jumper, 8% chose Morrison Soft TED, <1% chose Matagorda (Anon., 1988a)

October — Tested and submitted Parish TED for certification; Lettich TED failed; a Georgia and a Texas fisherman invited along on cruise (Anon., 1987; Christian and Harrington, 1987)

Brunswick—CES Fish and Wildlife Training Session (13 attended) (Anon., 1987)

ASMFC TED Seminar (professional meeting)—45 attended (Anon., 1987)

Workshops covering the installation and operation of the Morrison Soft TED were presented in 6 Florida ports by Marine Extension staff (Anon., 1987).

November — Key West—soft TED workshop (14 attended)
Through February 1988—remaining 23 pairs of TEDs distributed on first-come, first-served basis (Anon., 1988a)

December — Ft. Myers—soft TED workshop (8 attended)
St. Petersburg—soft TED workshop (11 attended)
Apalachicola—soft TED workshop (13 attended)
New Smyrna—soft TED workshop (13 attended)
Fernandina Beach—soft TED workshop (23 attended)

Louisiana Sea Grant

Center for Wetland Resources (Condrey and Day, 1987)

January — Through October—contracted services for three fishermen to conduct surveys of soft TEDs (had difficulty in locating individuals knowledgeable about TEDs, willing to come to Louisiana and share technology, considerable difficulty in locating and purchasing TED models from manufacturers)

Wanted to introduce TEDs to leading fishermen using individuals with first-hand knowledge; used series of contracts with fishermen to assist in general evaluation of the shrimp and bycatch retention patterns of various TEDs, and participate in series of workshops; enlisted SG Legal Program to address legal issues of NMFS mandate with fishermen

Four-state area SG—Louisiana, Texas, Mississippi, Alabama—involved in recruiting fishermen, conducting workshops on Morrison soft TED (reviewed legal issues, Endangered Species Act, NMFS congressional mandate), and demonstrating Georgia jumper; SG workshops held on sewing in new Morrison soft TED

Most significant problem: limited expertise in gulf on TED technology, making technology transfer difficult (relied upon private sector for assistance)

Mississippi-Alabama Sea Grant Consortium

TED workshops for net shops, fleet owners, and supply house operators floor samples of all certified TEDs; two TED designers available for questions; captains available who had used TEDs for shrimping; NMFS personnel to explain regulations—sponsored by SG Extension Service and Southeastern Fisheries Association (Force Five, 1987; Wallace and Hosking, 1988)

February — Gulfport, Mississippi
Biloxi, Mississippi
NMFS Pascagoula Lab
Bayou La Batre, Alabama (Wallace and Hosking, 1988)
Bon Secour, Alabama (Wallace and Hosking, 1988)

North Carolina Sea Grant

March — Sneads Ferry—Net and supply shop workshop (Bahen,

	1987a); discussed TED regulations and history of TEDs; covered Georgia Jumper, Hybrid, Stan-Mar, and NMFS TEDs construction and installation
	TEDs Fisherman's Forum (Bahen, 1987b)
	Morehead City—NMFS public hearing (100 attended) (Bahen, 1987b)
	Series of four workshops (travelling TED shows) held (Bahen, pers. comm., 1989; Bahen, 1987)
April	Beaufort, Bayboro, Wanchese, Varnamtown
	Delivered TEDs to two fishermen (Bahen, 1987b)
	Slide show on TEDs to Lion's Club (Bahen, 1987b)
May	Delivered one TED to fisherman (Bahen, 1987b)
	TED program New Hanover County museum (Bahen, 1987b)
	Met with NC SG communicators to outline plans for resuscitation poster to be published (Bahen, 1987b)
	Coordinated TED work at a Southeast Marine Advisory Network meeting (Bahen, 1987b)
	Delivered two TEDs to fishermen (Bahen, 1987b) Delivered TEDs to fishermen (Bahen, 1987b)
July	Participated in George's TED evaluation and testing of the soft TED (Bahen, 1987b)
	Delivered 12 TEDs to fishermen (Bahen, 1987b)
August	Began testing Morrison Soft TED and the Parrish Shooter Soft TED (Bahen, 1987b)
	Visited various fish houses to assess shrimp situations and give interested fishermen update on TED regulations; tried to demonstrate TED/discuss TED, try to get fishermen to try them (both accepted and asked to leave docks); selling point was elimination of small fish, crabs, jellyballs (Hart, 1987; Bahen, 1987b)
September	Working with net makers S. Parrish to redesign TED as a soft TED (Hart, 1987)

	Slide presentation to New Hanover County's Sierra Club (Bahen, 1987b)
	1-day trip to run final test on Parrish TED before Cape Canaveral trip (Bahen, 1987b)
October	Demonstration on TED devices during SG site review (Bahen, 1987b)
	Delivered Parrish TED to fishermen involved in finfish reduction project in Pamlico Sound (Bahen, 1987b)
November	Began planning for 1988 TED Purchase Project with DMF staff (Bahen, 1987b)

South Carolina Sea Grant

Displays on TEDs and sea turtles exhibited at Southeastern Wildlife Exposition (40,000 attendees nationwide) (Murphy, 1989)

Texas Sea Grant

Continued demonstrating TEDs; ran evaluations on Morrison soft TED (three boats) and Georgia jumper (nine boats); hired gear consultant to assist with tuning and TED installation

1988

Florida Sea Grant

Summer	Workshops conducted in conjunction with other marine agents—TEDs, covered NE Florida (Halusky, pers. comm., 1989)

Georgia Sea Grant

February	McClellanville, SC—TEDs workshop (20 attended) (Anon., 1988b)
	Kure Beach, NC—turtle workshop (24 attended)
	Augusta, Georgia—Georgia Fisheries Workers Association (27 attended)

	Richmond Hill, Georgia—Georgia Fisherman's Association (55 attended)
	Beaufort, South Carolina—South Carolina Shrimpers Association (90 attended)
	Brunswick, Georgia—Georgia Fishermen's Association (35 attended)
March	Baton Rouge—workboat seminar (429 attended)
	Charleston, South Carolina—South Carolina/Georgia Shrimpers Association, joint meeting (60 attended)
	TED presentation to Rotary Club (27 attended)
July	Eulonia, Georgia—Coastal Fisheries Advisory Meeting (47 attended) (Anon., 1988a,c)
August	Panama City, Florida—TED research (16 attended)
September	Tampa—Fish and Wildlife/CEE meeting (60 attended)
October	Atlanta—SESGMAS Regional TED meeting
	Galveston—NMFS meeting (7 attended) (Anon., 1988a)
	Eulonia, Georgia—Coastal Fisheries Advisory Meeting (135 attended)
December	Richmond Hill—Georgia Shrimpers Association meeting (46 attended)

Mississippi Sea Grant

January	Through March—tested MS Hybrid TED, transferred results to designer and SG agents (Anon., 1989c)
February	Conducted three TED technology transfer workshops with TED designers participating; fisherman hired as gear specialist—a 17-minute videotape created for Vietnamese (who produce 40% of the shrimp in Mississippi), describing TED regulations and TED types; copies left with two fleet owners employing majority of Vietnamese and at Catholic Social Services
	Bilingual Vietnamese conducted on-board demonstrations for Vietnamese and American fishermen

TEDs installed in 10 vessels (3 more also during year)—reports of rigging and use experience will be distributed to other fishermen and SG agents

Fishermen concerned that Coast Guard may overlook Vietnamese when enforcing TEDs correctly

April — Through June—Fisherman resigned as gear specialist (Anon., 1989c)

Two 30-minute television programs covering the impact of TEDs on fishermen and the Mississippi seafood industry

July — Through September—Tests conducted with NMFS to measure escapement rate of Kemp's ridleys from six certified TED designs, videos made (Anon., 1989c)

October — Through December—Meetings with representatives from SG and NMFS in Atlanta—plans to intensify educational efforts, although little can be accomplished until TED regs straightened out (Anon., 1989c)

Dockside visits, news releases explaining regulations

North Carolina Sea Grant

Because of uncertainty about implementation of TED regulations in 1988, the general consensus among Sea Grant directors was to "back off" during 1988, and North Carolina followed suit. The Sea Grant program had been under a lot of heat and there was a great deal of animosity towards them because of TED issue, and they were afraid of losing credibility. Bahen decided not to be as visible, but kept up with the TED regulation changes behind the scenes. Worked closely with the North Carolina Fisheries Association, dealt with television and newspapers frequently. Put in proposal and received an S-K grant to modify existing TED (Parrish and Parrish Grid) to reduce shrimp loss. Took improvements to MD and tested in flume, made full-scale TED and tested on *Georgia Bulldog* under shrimping conditions—purchased underwater camera, filmed capture and release of 100+ lb. turtle from TED and reduced shrimp loss to 6-10%, or less. (Modified TED has not been tested for certification—review panel not formed, no turtles at Canaveral; not sure but thinks it must be recertified; allowed North

Carolina fishermen to try it, no complaints (Bahen, pers. comm., September, 1989)

South Carolina Sea Grant

February — Through March—Charleston, Port Royal, McClellanville—three public hearings on the state TED regulations (Kohlsaat, 1988) (25 attended at Charleston, 21 at Port Royal, 5 at McClellanville (Murphy, 1989; Kohlsaat, 1988)

Personnel with Marine Resources Department held several workshops to demonstrate the various TEDs, but they were poorly attended by the industry (Murphy, 1989; Kohlsaat, 1988)

Texas Sea Grant

August — Through October—TED video completed: selecting TED types, tuning TEDs, advantages and disadvantages of various devices; ready for distribution (Graham, 1988)

NMFS TED Observer Program—provided Galveston Lab with needed gear expertise, TED tuning, educational opportunity to train captain and crew in TED handling and adjustments; TED installation techniques also demonstrated—one volunteer found

Continued working with fisherman hired as gear specialist/consultant

Attended regional Sea Grant directors meeting to discuss TED education activities

Participated in Brownsville/Port Isabel Shrimp Producers Association meeting—information regarding TED technology (few fishermen actually rigged for TEDs but interest in current technology exists)

Training session conducted for Texas County Marine Extension agents relative to TEDs, correcting areas where mistakes being made

Special education program for Gulf King Shrimp Co. in Aransas Pass, technology update

Fishermen rigging TEDs to exclude cannonball jellyfish;

	assisted 25 inshore and nearshore fishermen with TED technology
	Some offshore fishermen wanted to examine TED types without participating in NMFS Observer Program, after height of shrimp season
	Provided advice and information to potential TED manufacturers
	Assisted several inshore fishermen with installation of accelerator funnels (five demonstrations) Met with Texas Shrimp Association, Austin
November	Through December—More fishermen using TEDs to exclude large jellyfish (Graham, 1989a)
	Problems with homemade TEDs—did comparison tows, demonstrated proper devices and installation
	Demand for assistance with soft TEDs continued
	Evaluated problems with improperly modified Y-bridled lazylines with quad rings; evaluated differences between regular bridle array and that of extended bridles on standard rigs
	Vastly improved performance of Georgia jumper by lazyline modification—looking to see if improvements possible on standard trawls
	NMFS Observer Program continued with one participant

1989

Alabama Sea Grant

February	Through April—Arranged for updated Vietnamese translation of most current TED information times, areas, description of Parrish TED; updated lists of NMFS-approved TED manufacturers distributed to NMFS Regional Office, Vietnamese fishermen in Alabama and SG programs in five other states (Anon., 1989d)

Georgia Sea Grant

TED workshops and meetings (Anon., 1989a)

January	Tallahassee—Marine Fisheries Commission meeting (100 attended)
	New Orleans—NMFS TED meeting (34 attended)
	Tallahassee—Florida governor and cabinet turtle regulatory meeting (80 attended) Ft. Meyers—TED workshop (13 attended)
	Jekyll Island—Georgia Fisheries Workers meeting (38 attended)
February	Jekyll Island—Turtle Conservation Workshop (400 attended)
	Charleston—South Carolina Shrimpers Association meeting (85 attended)
	St. Petersburg—NMFS TED Workshop (12 attended)
March	Cape Canaveral—TED certification voyages on R/V *Georgia Bulldog* (20 attended, including several commercial fishermen from Georgia and other industry observers); looking for ways to reduce shrimp loss from Morrison soft TED; cruise to test Andrew TED; certified Morrison TED with Taylor-designed flap, Golden TED and Freeman TED, modified Morrison TED (cruise had one Georgia and one Mississippi commercial fisherman along) (Christian and Harrington, 1989; Anon., 1989a; Anon., 1989b)
	Morehead City—North Carolina Watermens Association meeting (45 attended)
	San Antonio—Texas Shrimp Association meeting (80 attended)
April	Through June—TED experience panel formed; made up of TED experienced commercial fisherman, NMFS and UGA MAREX specialists to pass on experience and technology to industry (Amos, 1989a)
	TED review panel formed of NMFS, commercial fishermen and UGA MAREX specialistis to review new industry-created TED designs and determine whether should be passed on to certification stage
	TED certification panel formed of NMFS and UGA

MAREX specialists, determine what methods can be used to get new TED design certified quickly

Tested Golden and Andrew TEDs in project in Panama City, Florida, in May; Morrison TED with exit hold flap was tested further

In conjunction with North Carolina Sea Grant, published installation instructions for Parrish TED

In contract report to GSAFDF—96 hours total towing time of control trawls and TEDs with performance and shrimp data collected for each tow (Amos, 1989b)

Two TV shows discussing TED effectiveness in saving turtles

Louisiana Sea Grant

June — Through September—VHS footage of TEDs in action above and below surface made available to SG agents and fishermen (Anon., 1989e)

Work done with individual fishermen regarding legality of their proposed design

Marine agents and specialists will coordinate info exchange among industry innovators, industry associations, net shops, federal labs, and other MAS programs in an effort to develop and improve TED Technology Transfer

Mississippi-Alabama Sea Grant Consortium

January — Through March—engaged respected shrimp fisherman as gear specialist to assist with TED technology transfer in MS: sit on industry review panel, daily hands-on operation of TED technology transfer on several vessels along northern Gulf Coast, assist Marine Resource Specialists in helping area fishermen to adapt more easily to regulations, assist fishermen in choosing and installing properly the device best suited for their operation (Veal, 1989)

Developed list of potential TED sale outlets and distributed them to TED manufacturers (Anon., 1989c)

March	Participated in *Georgia Bulldog* cruise in Cape Canaveral shipping channel—soft TED certification proceedings (aborted due to scarcity of turtles) (Veal, 1989)
	Attended meeting of Concerned Shrimpers of America, Biloxi (Veal, 1989) (Anon., 1989c)
	Gear specialist (fisherman) resigned
May	Through July—Further attempts made to involve Vietnamese in at least recognizing that TED regulations do exist (Anon., 1989c)

North Carolina Sea Grant

Submitted proposal to SAFDF to help industry with on-board tuning of TEDs to make their use as efficient as possible. Contacted a fisherman to work with them who was familiar with TEDs, helped fishermen sew in TEDs, tune the nets, etc. The changes with the TED regulations implementation and enforcement in 1989 meant the fishermen removed the TEDs from nets. Spent a lot of time with media. In September, fishermen called to complain that Coast Guard was checking for TEDs; evidently, the old regulations were back in complete effect (however, the season for North Carolina ends August 31 for TED use); had to update and inform Coast Guard about regulations. Informed North Carolina fishermen that as they follow shrimp south to Florida, must be aware and comply with Florida TED regulations (year-round requirement for TEDs); believes that there's almost 100% compliance with TED regulations in Florida and Texas; Louisiana is the problem state at this time (pers. comm., Bahen, September, 1989)

Texas Sea Grant

February	Through April—two evaluations conducted with modified Georgia jumper (Graham, 1989)
	Advised company developing fiberglass jumpers (proper installation techniques demonstrated, net person trained, offshore performance evaluation (Graham, 1989)
	Several demonstrations conducted for net makers regarding construction of Morrison soft TEDs—on-board demonstrations given for seafood companies and small groups of fishermen (cautioned against their use due to

shrimp losses, but still preferred by fishermen for safety and cost factors) (Graham, 1989)

Directed efforts to continue assisting fishermen in proper installation and construction of TEDs; problems with faulty homemade jumpers (Graham, 1989)

No cooperators found for this quarter for NMFS TED observer program, several companies interested for later in year (Graham, 1989)

Rigged boat with two Sanders Grid TEDs and two Boone jumpers—problems with shrimp loss corrected by fisherman/gear specialist (Graham, 1989b)

TEDs adapted to exclude large jellyfish

Updated TEDs fact sheet

Appendix I

Newsletters and Notices Pertaining to TEDs

North Carolina Sea Grant

1981

Coastwatch (1989 circulation 22,000, general audience, ten times a year)

May: describes NMFS emergency 240-day regulations for reviving comatose turtles (resuscitation)

1988

Coastwatch

February: notice of certification of Parrish TED, including description and cost; notice of availability of DMF money for the TED purchase plan to be presented at 6 public meetings; information presented on different kinds of TEDs; summarizes performances; outlines how to apply for money to buy certified TED (North Carolina Department of Commerce's Energy Division)

1989

Coastwatch

March: article describes designing, testing, and certification of Parrish soft TED; proposes future modifications; notes that most NC shrimpers have not bought a TED, hoping TED regulations will be abolished

---

Compiled by Nancy Balcom and William DuPaul, Virginia Institute of Marine Science, College of William and Mary.

Marine Advisory News (1989 circulation 2200, quarterly)

January/February: published instructions for applying for reimbursement for purchase of TEDs up to $500, must meet certain conditions

Legal Tides (1989 circulation 630, quarterly)

Articles discuss history of TEDs issue regarding legality, overview of pending lawsuits

North Carolina Tar Heel Coast

May: describes final TED regulations effective May 1-August 31, stating that if there is not 80% compliance with the regulations in 1989, regulations will be effective April-September in 1990; provides overview history of TED development, describes types available, and reiterates the regulations; gives penalties for noncompliance with regulations and violation of Endangered Species Act; notes warnings will be given until July 1 on a case-by-case basis

Florida Sea Grant

1987

October: list of TED manufacturers

1989

The Marine Times (1989 circulation 545, every 2 months, general public and commercial fishermen)

May-June: notes regulations scheduled to go back into effect May 1, 1989 requiring shrimp trawlers to use TEDs and 90-minute tows in offshore waters (regulations apply to all U.S. offshore waters from North Carolina to Texas; describes regulations in detail; covers emergency regulations in effect in Florida; and describes 6 certified TEDs

Alabama Sea Grant

1983

Sea Harvest News (1989 circulation 1700 to Alabama, 200 to Mississippi; monthly; commercially oriented)

November: reports on development of devices to reduce incidental catches of turtles caught in shrimp trawls, reduce bycatch of finfish; describes cruise on R/V *Oregon* (many fishermen on east coast and in Los Angeles are using TEDs with good success); notice of upcoming January/February workshops

1985

Sea Harvest News

February: announces and describes NMFS TED and its benefits in reducing bycatch

1986

Sea Harvest News

January: lists intentions of new proposed regulations (Phases I and II, net exemptions)

March: notice to shrimpers on request from U.S. Fish and Wildlife Service and conservation groups to Gulf of Mexico Fisheries Management Council to implement mandatory TED regulations; lists availability of brochure ("Construction and Installation Instructions for the Trawling Efficiency Device")

November: describes proposed mandatory TED regulations, tests on the NMFS TED, the mediation process

1987

Sea Harvest News

January: notes approach of January 1987 deadline for using TEDs in southern Florida area; describes 5 approved TED designs; lists availability of drawings and installation instructions; updates regulations for southern Florida (136)

April: update on mediation talks

June: updates forthcoming (July 15) regulations; reports findings of commercial shrimp boat working Tampa to Key West, testing both NMFS TED and Georgia Jumper

July: provides information on regulations for Canaveral area (emergency regulations effective October 1); updates rest of upcoming TED regulations (1988)

September: updated Morrison soft TED certification

October: introduces availability of Morrison soft TED; covers shrimp catch tests on Georgia Bulldog (14% loss brown shrimp, 5%

loss white shrimp); describes boardings conducted by NMFS agents in 2 Florida ports after emergency regulations went into effect)

1988

Sea Harvest News

January: presents TED regulations in full

March: TED questions and answers taken from questions that have come up at TED meetings in Alabama, Mississippi, and Louisiana

June: notice to fishermen of TED appeal from Concerned Shrimpers of Los Angeles; notes injunction against enforcing TEDs remains in effect

1989

February-April—progress report: continue to maintain low profile in dissemination program, with activities primarily confined to providing responses to questions from the industry relating to the resumption of TED regulations (most Alabama shrimpers do not believe regulations will be reinstated); update and revise printed material prepared for use by Alabama shrimpers, including the most current information on times and areas requiring the use of TEDs, a description of the Parrish TED, and an updated listing of manufacturers of TEDS approved by NMFS; remain ready to respond to requests from shrimping industry for information on both TED regulations and construction (although availability of information has been announced, there has been an extremely limited number of requests for it); have arranged for an updated Vietnamese translation of the most current TED information on times and areas, etc., as above (publication distributed to NMFS, Regional Office, Vietnamese shrimpers in Alabama, and Sea Grant programs in the other gulf states)

Sea Harvest News

August: notice to fishermen that TED regulations for offshore waters will go into effect May 1, 1989 and for inshore waters May 1, 1990

Mississippi-Alabama Sea Grant Consortium

1987

Force Five (1989 circulation 1200; quarterly; educators, general public, marine resource managers, university faculty, some commercial fishermen)

Winter: announcement of publication "Turtles, Trawlers, and TEDS: what happens when the Endangered Species Act conflicts with fishermen's interests," an in-depth look at the TEDS controversy printed in Water Log (Sea Grant's legal reporter); issues also contains replies by Mike Weber and T.J. Mialjevich

1988

Booklet describes TED regulations, certified TEDs with diagrams, locations of where 90-minute tows and TEDs are required, including excerpts from the *Federal Register*; booklet also translated into Vietnamese

Mississippi Sea Grant

1987

Gulf Coast Fisherman (1989 circulation 1500, monthly, commercially oriented)

January: TED regulations update; proposed regulations with 1-$^1/_2$-year phase-in period, with different seasons and different geographical coverage, depending on shrimp statistical zone; provides source for detailed explanation and summary of proposed regulation

July: summary of final regulations regarding TEDS as published in detail in the June 29 *Federal Register*

November: notice of certification of Morrison soft TED effective October 1; explains redefined inshore and offshore waters corresponding to the 1972 International Regulations for Preventing Collisions at Sea (72COLREGS) demarcation line; provides source for details on TED and the regulations; reports on NMFS dockside boardings in Cape Canaveral area, with at-sea boardings to begin in November

1988

Gulf Coast Fisherman

May: notice of enjoinment of NMFS from enforcing TED regulations pending appeal

August: notes TED regulations back in effect September 1; describes TED regulations in full; describes possibility of Heflin amendment

October: reports TED regulations back in effect September 1, warnings issued until November 1, Congress acting on Heflin amendment; lists sources of information on where to obtain TEDs and proper installation techniques

November: reports bill signed reauthorizing the Endangered Species Act to 1992, along with Heflin amendment delaying implementation of TED regulations until May 1, 1989; reviews regulations

1989

Gulf Coast Fisherman

January: reports record number of ridleys washed up on the Georgia/Florida coast during fall 1988; Florida Marine Fisheries Commission enacting emergency year-around regulations for northeast Florida requiring the use of TEDs; notes information available from Sea Grant (call or write)

"Mississippi Hybrid TED" (advertising brochure)—provides picture, description, and availability

January-March—quarterly progress report: reports that a list of potential outlets for TEDs has been developed and forwarded to manufacturers of two types of TEDs (information available to any manufacturer)

Louisiana Sea Grant

1987

Report summarizes rules and regulations designed to protect sea turtles as listed in the *Federal Register* (1987); describes regulations, certified TED types and their installation, results of tests on various TEDs

St. Bernard Fisherman (Cooperative Extension)

Spring newsletter: Q&A format concerning the sea turtle situation, the Endangered Species Act, TEDs, and enforcement; reports proposed regulations; urges awareness of shrimpers to the situation ("won't just go away")

1989

January-June—interim progress report: due to delay in implementation of regulations, there has been little or no progress on work scheduled; most shrimpers view the suspension as a victory, therefore little education impact has been possible, other than keeping shrimpers and local legislators informed of the status of TED regulations; publish regular newsletter, schedule meetings, and visit with individual shrimpers

Cajun Courier (Cooperative Extension)

June: reported regulations concerning TEDS scheduled for full enforcement July 1, 1989 (TED law officially in effect May 1): brief overview of regulations; warnings issued through end of July, except in cases of repeated or flagrant violations; list of commonly asked TED questions and answers

Louisiana Shrimp Association

The Net Log

Discussion of meeting held in Washington, D.C., in August 1986, between shrimp industry, environmental groups, and NMFS/NOAA; proposal on regulations effective May 1987, showed NMFS data on turtle incidental catches by shrimpers

Georgia Sea Grant

1987

July-September—quarterly report: upon submission of Morrison soft TED for certification, subjected it to shrimp and biomass retention tests; arranged for testing of downward-shooting TED in October

1988 *Georgia Jumper Installation Instructions*—diagrams and instructions

Morrison Soft TED Installation Instructions—diagrams and instructions

Parrish Soft TED Installation Instructions—diagrams and instructions

The Salty Dog

November: lists availability of publications giving installation instructions for Georgia Jumper and Morrison soft TED

1989

The Salty Dog (1989 circulation 1300, quarterly, commercially oriented)

February: lists availability of installation instructions for Georgia Jumper and Morrison soft TED

May: notice about availability of installation instructions for Georgia Jumper and Morrison soft TED

Sea Turtles in Georgia (brochure in *Coastcards* series): reports that the Office of Energy Resource of the Department of Natural Resources purchased and distributed more than 800 TEDs to Georgia shrimpers

South Carolina Wildlife and Marine Resources

1986

December—news release: describes tests of NMFS TED and Georgia Jumper; compares shrimp retention/loss and retention/reduction of finfish, blue crab, and horseshoe crab by catch between standard nets and nets with TEDs installed; purpose of tests is to give shrimpers choice of TEDs should their use become mandatory

1989

June: personal letter describing department news releases, newspaper articles, editorials, and letters to the editor; summarizes 1988 (notes that hardly a day went by during the summer without some coverage of TEDs or turtles in the news, including television and radio coverage)

Texas Sea Grant

1988

February: provides information sheet giving answers to commonly-asked questions about TEDs, regulations, enforcement, building and installation, shrimp retention, distances offshore, net sizes, and incidental turtle catches; gives list of contacts for information and technical assistance

Center for Marine Conservation

1982

February—report: preliminary estimate of the payoff to investing in a turtle excluder device for shrimp trawls

1989

May—report for the Florida Marine Fisheries Commission: telephone survey of TED manufacturers determined that many were available for purchase, however few sold. Fishermen waiting to see if fines issued; enforcement agencies and shrimpers indicated that TED availability is limited (contradicted survey of TED manufacturers); enforcement of regulations spotty, but picking up as a result of public and conservation pressures; additional enforcement problems typically center around various ways to appear to be pulling TEDs, while actually not having functioning TEDs

National Marine Fisheries Service

1984

NMFS Newsbreaker

No date—special edition on sea turtles: describes the 8 species of turtles and the management and research program ongoing for turtles, including TED technology transfer, TED development and testing, different sources of mortality, etc.

1985

June: NMFS booklet describes how to build, install, and handle NMFS TED; describes benefit to using device

1988

November: summary of tow-time and TEDs regulations by area, giving starting date, season, and vessel requirements

1989

March: news release covering TED regulations going into effect May 1; lists regulations, emergency regulations already in effect, the names of the 6 certified TEDs, where to send for copies of the regulations, and a list of TED manufacturers

May: list of TED manufacturers

Date unknown

Describes turtle resuscitation procedures for shrimpers; discusses preserving a livelihood, use of TEDs, and NMFS funding of TEDs; promotes voluntary use

"Enforcement of Turtle Excluder Device:" list of 9 commonly-asked questions and answers regarding TED enforcement

Washington, D.C.

1989

Marine Fish Management

March: notice published about TED and tow-time regulations about to go into effect May 1, 1989: lists where copies of regulations available; covers emergency regulations in effect in northern Florida and southern Georgia

South Atlantic Fishery Management Council

1989

May: describes TED and tow-time regulations in effect and emergency regulations in effect in southern Georgia and northern Florida; lists 6 certified TEDs; gives address and number to obtain TED regulations, list of TED manufacturers

Biographies of Committee Members

JOHN J. MAGNUSON, *Chairman*, is a professor of zoology and director of the Center for Limnology at the University of Wisconsin in Madison. He received a B.S. in 1956 and an M.S. in 1958 from the University of Minnesota, and a Ph.D. from the University of British Columbia in 1961. He served on the National Research Council's Ocean Policy Committee in 1979. He has also served as ecology program director at the National Science Foundation, as president of the American Fisheries Society, and scientific adviser to the Great Lakes Fishery Commission. Dr. Magnuson's research interests include behavioral ecology of fishes, distributional ecology of fishes and macroinvertebrates, and lake community structure. He is currently the principal investigator of the Long-term Ecological Research site on northern lake ecosystems.

KAREN A. BJORNDAL is an assistant professor of zoology and director of the Archie Carr Center for Sea Turtle Research at the University of Florida in Gainesville. She received a B.A. at Occidental College in 1972, and a Ph.D. from the University of Florida in 1979. She served on the Marine Turtle Specialist Group of the International Union for the Conservation of Nature and as marine turtle conservation coordinator for the World Wildlife Fund U.S.A. She is currently a research fellow of the Caribbean Conservation Corporation and a member of the U.S. Recovery

Team for Green Turtles and Loggerheads. Dr. Bjorndal's research includes sea turtle demographics, feeding ecology, and growth rates and nutrition.

WILLIAM D. DUPAUL is the chairman of the Department of Marine Advisory Services, Virginia Institute of Marine Sciences, College of William and Mary, at Gloucester Point, Virginia. He received a B.S. from Bridgewater State College in 1965, and an M.A. in 1968 and a Ph.D. in 1972 from the College of William and Mary. He was a commercial lobster and scallop fisherman from 1975 to 1977, and he also worked with the Gulf and South Atlantic Fisheries Development Foundation and Sea Grant in the Mid-Atlantic Region. Dr. DuPaul's research includes commercial fisheries development, fisheries development, fisheries management, and marine resources.

GARY L. GRAHAM is a marine fisheries specialist with the Sea Grant College Program, Texas Agricultural Extension Service, Freeport, Texas. He received a B.S. from Texas A & M University in 1969. He has served on the advisory committee to the Texas Shrimp Development and Research Foundation and on the Reef Fish Advisory Committee, Gulf of Mexico Fisheries Management Council. For five years Mr. Graham was a commercial shrimp fisherman in the Gulf of Mexico and has experience in commercial culture of marine shrimp in ponds. Mr. Graham's research interests center around fish and shrimp harvesting equipment, shark fishing, and loran obstructions.

DAVID W. OWENS is a professor of biology at Texas A & M University in College Station, Texas. He received a B.A. from William Jewell College in 1968 and a Ph.D. from the University of Arizona. He has served as team leader of the Kemp's Ridley Recovery Team, participated in numerous sea turtle conferences, workshops, and symposia, and served on the editorial board of *Marine Turtle Newsletter.* His research centers on endocrine control of reproduction in sea turtles and sea turtle physiology. He is the author of numerous scientific articles and reports on sea turtles.

CHARLES H. PETERSON is a professor of marine science at the University of North Carolina at Chapel Hill. He received a B.A. from Princeton University in 1968: and an M.A. in zoology in 1970, and a Ph.D. in biology in 1972 from the University of California at Santa Barbara. He was a member of the National Research Council's Committee on Options for Preserving Cape Hatteras Lighthouse. He has served on a biological oceanography panel for the National Science Foundation and on the

North Carolina Marine Fisheries Commission. Dr. Peterson's chief areas of research include population biology and community ecology of marine benthic invertebrates, life history patterns, and fisheries management.

PETER C.H. PRITCHARD is vice president for science and research at the Florida Audubon Society in Maitland, Florida, where he has been since 1973. He received a B.A. in 1963 and an M.A. in 1968 from Oxford University, and a Ph.D. in 1969 from the University of Florida in Gainesville. He has worked as a turtle specialist for the World Wildlife Fund and International Union for the Conservation of Nature. His extensive international experiences with endangered sea turtles have ranged throughout Latin America and the Pacific region. He is the author of many scientific articles and reports and several books including *Encyclopedia of Turtles* and *The Turtles of Venezuela.*

JAMES I. RICHARDSON is director of the Georgia Sea Turtles Cooperative Research and Education Program at the Institute of Ecology in Athens, Georgia. He received a B.S. from Juniata College in 1965, and an M.S. in 1970 and Ph.D. in 1982 from the University of Georgia. He has served on numerous boards, panels, and advisory committees focusing on worldwide and regional sea turtle conservation issues. He and his coworkers have authored many reports and scientific publications on sea turtle population biology. Dr. Richardson's principal research interests are on conservation and ecology of sea turtles, endangered species populations, and resource management.

GARY E. SAUL is an environmental scientist with FTN Associates in Little Rock, Arkansas, but in the course of writing this report, he became an assistant professor of biology at Southwest Texas State University in San Marcos. He received a B.S. at North Carolina State University in 1972, an M.S. from Louisiana State University in 1977, and a Ph.D. at Virginia Polytechnic Institute and State University in Blacksburg in 1980. For the Texas Parks and Wildlife Department, he has served as director of fisheries harvest programs and finfish programs. Dr. Saul's research has involved fisheries management, harvest statistics, and fisheries resources. He has participated in numerous seminars and panels dealing with marine resource management.

CHARLES W. WEST is manager of research and development at Nor'Eastern Travel Systems Inc. in Bainbridge Island, Washington. He received a B.S. in 1979 and an M.S. in 1985 in fisheries at the University of Washington in Seattle. He has served with the National Marine Fisheries Service's Resource Assessment and Conservation Engineering Division of the Northwest and Alaska Fisheries Center. He holds a patent for a sorting device used in trawl nets. Mr. West is active in the Marine Technology Society and the International Council for the Exploration of the Sea. His research and development interests include design of trawling systems and fish behavior and fish/gear interactions.

Index

G

H

I

K

L

M